Philosophy and Technology

ROYAL INSTITUTE OF PHILOSOPHY SUPPLEMENT: 38

EDITED BY

Roger Fellows

CAMBRIDGE
UNIVERSITY PRESS

Published by the Press Syndicate of the University of Cambridge
The Pitt Building, Trumpington Street, Cambridge, CB2 1RP
40 West 20th Street, New York, NY 10011-4211, USA
10 Stamford Road, Oakleigh, Melbourne 3166, Australia

© The Royal Institute of Philosophy and the contributors 1995

*A catalogue record for this book is available
from the British Library*

Library of Congress Cataloguing in Publication Data

Philosophy and technology/edited by Roger Fellows
    p. cm.—(Royal Institute of Philosophy supplement: 38)
       1. Technology—Philosophy. I. Fellows, Roger. II. Series
  T14.P497 1995
  601—dc20   95–18543
                                         CIP

ISBN 0 521 55816 6

Origination by Michael Heath Ltd, Reigate, Surrey
Printed in Great Britain by the University Press, Cambridge

Contents

Contents

Introduction

ROGER FELLOWS

The essays collected here do not constitute a philosophy of technology, in the sense which, for instance, Don Ihde requires.
According to Ihde the philosopher of technology must reflectively
analyse technology in such a way 'as to illuminate features of the
phenomenon of technology itself'.[1] The contributors to this volume do not concern themselves with the essentialist enterprise of
defining technology; they more or less take it for granted that the
reader is familiar with a variety of technologies such as
Information Technology, and proceed from there. Hence the title
is the conjunctive one of *Philosophy and Technology*.

That contemporary philosophy has become more self-consciously concerned with the impact of technology on human
nature and society is undeniably. Witness for instance the philosophical growth areas of Environmental and Medical Ethics. This
interest is surely in part a consequence of environmental disasters
such as the Chernobyl nuclear accident, and advances in medical
technology such as organ transplantation. Again, advances in computer science and technology have suggested new ways for self-
understanding and the re-organisation of society.

The modern world contains a vast array of technologies, and the
contributors to this book respond to some of them with different
concerns, so there is no one underlying theme running throughout. Nevertheless, one may discern that, in general, the contributors adopt one of two approaches. The first is concerned, quite
generally, with the impact of technology on culture and society,
and the second is concerned with philosophical questions raised by
particular technologies. The papers by David Cooper and Stephen
Clark exemplify the former approach, and the contributions by
Willem Hackmann and Sophie Botros the latter.

David Cooper addresses the question of whether technology is a
force for liberation or enslavement. He disentangles various issues
connected with liberationist and enslavement claims about contemporary technology. His main conclusion is that Western tech-

---

* The essays in this volume derive from papers delivered at a Royal
Institute of Philosophy conference held at the University of Bradford in
July 1994, on the theme of Philosophy and Technology.

[1] Don Ihde, *Philosophy of Technology* (New York: Paragon House,
1987), p. 38.

1

nological societies have eroded the notion of the self conceived of as an autonomous entity, enmeshed in a system of rights and responsibilities. The issues raised by Cooper are important: consider for instance IT, and, in particular, E-mail and the Internet. On the one hand, as Cooper notes, it might be thought that we are possessed of a technology which enhances our freedom by enabling us, for instance, to 'E-mail' our objections to, or support for, a certain policy directly to 'Government'. But, on the other hand, Cooper rightly enquires whether a Government which responded to 'grouses and grumblings' on E-mail would be behaving responsibility. The problem here has to do with a severance between a technological ethos ('here is a computer which enables you to complain to the centre'), and a sense of the moral, the political and the aesthetic about which E-mail in particular, and IT in general, are silent. IT does not help in giving the citizen the intellectual and moral groundings which are the necessary conditions for its humane use. It may even mitigate against the inculcation of these virtues, because of the solipsistic potential inherent in a work force of persons tapping away at computer keyboards in their own homes.

The papers by Smithurst, Hackmann, Hendry and Cartwright are all concerned, in different ways, with the relations between science, technology and reality. Smithurst discusses the question as to whether successful technologies confirm the truth of scientific theories, and failed technologies refute them. We might think, for instance, that our ability to send a spacecraft to the moon confirms classical mechanics. Smithurst argues that the relationship between our technologies and our theories is not like that between observation and theory. Imagine that all the scientific instruments developed since the seventeenth century were to vanish overnight. Then, says Smithurst, our scientific theories would rapidly degenerate into metaphysical myths with the same status as the atomic theories of Leucippus and Democritus.

Hackmann focuses on the role of instruments in the study of nature. He provides an illuminating and detailed case study of our understanding of the phenomenon of the aurora borealis by concentrating upon the interplay between theory and experimental apparatus. Like Smithurst, Hackmann sees an indissoluble link between theory and technology, and he ends his paper with a conceptual tension produced by the history of scientific instruments. As we become ever more reliant on scientific instruments in the investigation of nature, so do we become increasingly more estranged from nature. Bertrand Russell quipped that naive realism leads to physics and the truth of physics entails the falsity of naive realism. Russell had in mind for instance the point that our common sense ideas about simultaneity are overthrown by Special Relativity theory. Hackmann's point is rather that we study the world behind a barrier of instruments, which serve to insulate us from nature.

In his contribution, Hendry urges that scientists should not be instrumentalists, but ought rather to adopt the position of the methodological realist: if our most relevant theory posits the existence of quarks then we should believe that quarks exist. For the scientific realist there is a way that the world is, and it is the aim of science to discover it. Cartwright on the other hand is a pluralist. If it is true that the world is just one way, then the Quantum Measurement Problem requires that all true descriptions of reality are renderable as Quantum descriptions. But she argues that we do not need to choose between quantum mechanical descriptions and classical ones. The methodological realist is committed to providing a theory of the relation between quantum and classical states, or at least to believing that such a theory exists. Cartwright sees in this kind of wholesale imperialism and reductionism. She argues that there are both quantum states and classical states, and that there is no contradiction between them. For some scientific ends we invoke one set of properties; at other times, others.

Fellows discusses some aspects of the question as to whether a computer could be endowed with genuine mentality. The computer is a technology which has become embedded in popular culture, to say nothing of its influence in the disciplines of philosophy and psychology. Fellows' discussion is mainly focused on the work of the American philosopher John Searle. Searle is the most influential critic of theorists of artificial intelligence who hold the view that a digital computer could think provided only that it was properly programmed, and of philosophers and psychologists who maintain that the computer provides the best means of understanding ourselves.

There are as well moral dilemmas involved in the construction of thinking machines, but many will think that they lie in the distant future. However, the moral problems thrown up by modern medical technologies are very much with us, and we have to try and solve them within our existing moral frameworks or to evolve a 'new ethic'. Botros provides a sensitive discussion of one dilemma brought about by medical technology. Until recently, persons who sustained massive damage to the cerebral cortex could not have survived the initial trauma which caused it. But given that the brain stem is intact, so that he or she can breath unaided, the patient can be kept alive for years by being artificially fed and hydrated. Such patients are not clinically dead, but their chances of recovery are nil. Ought doctors to sustain the lives of such patients, or to withdraw the technology with the result that the patient will die of starvation?

Gyekye is a Ghanian philosopher who has for a long while been interested in African thought. In his paper, he stresses the empirical orientation of African thought as exemplified in agriculture and herbal medicine. One might have supposed that such an outlook would have led naturally to an interest in theoretical science; for instance, Egyptian rules of thumb for measuring the areas of fields led ultimately to Euclid's axiomatisation of plane geometry. However, there is no evidence that this was the case. Gyekye discusses why the explosive coruscation of achievement which is Western science and technology never occurred in Africa.

O'Hear, in his paper, points out that it might appear that modern technology, such as computer-driven graphics and new materials, offers dramatic possibilities for artistic expression. He believes, however, that technological advancement is, on the contrary, affecting art adversely. Suppose we discovered that what we had thought to be an original work of art, say a picture, has been computer-generated: ought we to continue to regard as a work of art? O'Hear thinks not. Part of his argument relies upon a form 'externalism', according to which works of art are constituted not merely by their visible forms, but also by the fact that they have been intentionally produced in a particular way to produce a certain kind of response in their audience.

The concerns of Clark and Grant overlap with those of Cooper. Our technologies are becoming increasingly complex, and Clark suggests that we may be on the brink of an era when no one will really understand the machines which run our lives. Hence the idea that human societies will move ever closer to being rationally controlled by computers and other forms of technology may be an illusory one. It is not just that we could not understand the controlling machines because of their complexity; but rather that the more complex computers became, the more they would be like living things and thus subject to the same evolutionary twists and turns as all living things.

Technology treats the natural world as a means to human ends, and Grant has no quarrel with that. He is worried, however, about Technocracy, that is rule by technicians. Grant is right to be concerned. The results of the new technocratic attitude to human beings and their affairs lie all around us, and are manifested in the rise of the new managerialism with its emphasis on measurement (to say nothing of the odd belief that one has somehow created a different order of things in universities or hospitals by employing a firm of consultants to design a new logo).

Renford Bambrough has now retired as editor of the journal *Philosophy*, and so it is fitting to include a paper by him reflecting upon the practice of philosophy. Bambrough believes that the philosopher ought, wherever possible, to avoid scientific and technical jargon, and to present his or her thoughts in plain language. he diagnoses the tendency of some philosophers to retreat into technocratic jargon as a consequence of the fear of being thought to be 'too literary', and not scientific enough. But perhaps a defence against this tendency would be a realisation that technique is not the same as rigour.

I should like to end this introduction on a personal note. Bradford is primarily a technological university, and it gave me great pleasure to co-organise a conference on philosophy and technology there. Many of the participants commented on how cooperative and relaxed the proceedings at the conference were, and my hope is that it may be possible, in the light of the conference, to get together a group of philosophers and technologists interested in exploring further the impact of technology on society.

Technology: Liberation or Enslavement?

DAVID E. COOPER

I

The week, twenty-five years ago, of the Apollo spacecraft's return visit to the moon was described by Richard Nixon as the greatest since the Creation. Across the Atlantic, a French Academician judged the same event to matter less than the discovery of a lost etching by Daumier. Attitudes to technological achievement, then, differ. And they always have. Chuang-Tzu, over 2,000 years ago, relates an exchange between a Confucian passer-by and a Taoist gardener watering vegetables with a bucket drawn from a well. 'Don't you know that there is a machine with which 100 beds are easily watered in a day?'—'How does it work?'—'It's a counter-balanced ladle'—'Too clever to be good ... all machines have to do with formulae, artificiality [which] destroy native ingenuity ... and prevent the Tao from residing peacefully in one's heart'.[1] 'Engines of mischief', in the words of the Luddite song, or testaments to 'the nobility of man [as] the conqueror of matter',[2] in those of Primo Levi, the products of technology continue to inspire phobia and philia.

Familiar to the point of cliché, in such debates, has been the rhetoric of liberation and enslavement. Karl Popper judges nuclear power to have shown that our vaunted 'control of nature' is 'apt to enslave us rather than make us free';[3] while, for Radhakrishnan, further east, the technology by which man 'strove to emancipate himself from bondage to nature' has now become 'the master'.[4] Both statements, of course, sound the familiar thought that what was, ideally or originally, a force for liberation has, in subsequent practice, become the opposite.

Chuang-Tzu's ladle or Radhakrishnan's inventions that protected stone-age man from raw nature will, for some, be insufficiently sophisticated or informed by scientific theory to count as technology. I am not going to agonize over the 'proper' definition

[1] Zhuang Zi, in *Wisdom of the Daoist Masters* (Lampeter: Llanerch, 1984) Sect. 12K.
[2] *The Periodic Table* (London: Abacus, 1984), p. 42.
[3] *Realism and the Aim of Science* (London: Hutchinson, 1983), p. 260.
[4] 'Fragments of a confession', in P. Schilpp (ed.), *The Philosophy of Sarvepalli Radhakrishnan* (New York: Tudor, 1952), p. 19.

David E. Cooper

of 'technology', if such there be, though it is easy to recognize some candidates as too narrow or too broad. 'The manipulation of the environment in the interests of human life',[5] for example, hardly covers information technology; while 'the practical implementation of intelligence'[6] would cover running for shelter from a storm, scarcely a technological feat. I shall not try to improve on these efforts, and do not need to, for everyone will accept as paradigms of the technological enterprise the practices or processes that concern me—industrialization, computer technology, modern medicine, genetic engineering. To be sure, there are interesting questions about, if you will, the 'essence' of technology, but they are not, except in passing, mine.

My concerns are, first, to lend order to the welter of claims pro and con the technological enterprise which are couched in the rhetoric of freedom and enslavement; then to say a little about the relations between these claims; and, finally, to elaborate on what is, perhaps, the most philosophically charged theme to emerge from the confrontations described.

II

Let me begin in the liberation camp. What do those who speak of technology as a force for freedom have in mind? Their many claims fall, at a pinch, into three kinds, beginning with ones to the effect that technology confers various 'freedoms from'. Well, from what? Most obviously, from Radhakrishnan's 'bondage to nature'. The light-bulb and the aeroplane free us, to a degree, from the constraints of darkness and distance. But, especially with the arrival of IT, other 'freedoms from' are credited to technology: from ignorance and prejudice, for example. The weather satellite tells the sailor of next week's storm; the data bank quickly supplies the figures that biased parties would otherwise make up. During the 1970s, Salvador Allendé briefly entrusted policy to a giant computer designed by a British engineer who described it as 'the Liberty Machine'. He did so because '"Liberty" may be redefined for our technological era' as 'competent information is free to act'—free, that is, from bias, sentiment and guesswork.[7]

[5] Ibid.

[6] Frederick Ferré, *Philosophy of Technology* (Englewood Cliffs: Prentice Hall, 1988), p. 33. Ferré does specify the terms in his definition to avoid such a counter-example.

[7] Quoted in Theodore Roszak, *The Cult of Information* (Berkeley: University of California Press, 1994), p. 225.

Operative in these and comparable claims is, perhaps, the old Stoic idea of freedom as invulnerability to whatever threatens the rational self. For Epictetus, this required retreat from the world—a thoroughly threatening place—into one's inner sanctum. But for today's Stoic, there is the alternative of invulnerability to threats through technological pre-emptive strikes. Whereas for Epictetus, and also the Epicureans, the crippling fear of premature death was avoided by coming to view death as 'nothing', today we can rely on medicine virtually to ensure that it won't be premature. Technology, muses a Don DeLillo character, is what we have invented to 'conceal the terrible secret of our decaying bodies'.[8]

A second set of claims registers a more 'positive', Promethean notion of freedom. It was this which, earlier in our century, inspired enthusiasm for technology among Italian Futurists, like Marinetti, and the so-called 'reactionary modernists' of the German Right, such as Oswald Spengler.[9] The technologist, the 'conqueror of matter', vividly embodies that will-to-power which, for Spengler, induces 'man to pit himself against' nature. In technological or 'Faustian man' is realized 'the destiny of the free personality', which is 'to build a world oneself, to be oneself God'.[10]

This freedom as active will-to-power is not the mere flip-side to freedom from the vicissitudes of nature. Faustian man, insists Spengler, 'cares not a jot' whether the products of his technology are 'useful', in elongating our lives, say, or protecting us from the weather. The analogy is with the artist—the potter, say, whose creative freedom resides in the imposition of form upon materials, and not at all in the storage potential of the pots to reduce dependence on a good rainfall or harvest.

Spengler's favoured examples of technology were the foundry and the shipyard, where Faustian men, like so many Nibelungen, hammered the world into shape. More recent technology affords new scope for Promethean rhetoric. 'You can create your own universe!', proclaims one hacker: and with the added *frisson* that you are controlling an 'intelligent' machine, not a mere lump of metal. 'You could control [the computer]', says another hacker, 'you could be God', subjecting its 'smartness' to yours.[11] More satisfy-

---

[8] *White Noise* (London: Picador, 1986), p. 285.

[9] The term is Jeffrey Herf's. See Michael F. Zimmermann, *Heidegger's Confrontation with Modernity* (Bloomington: University of Indiana Press, 1990), pp. 47ff.

[10] *Man and Technics: A contribution to a philosophy of life* (London: Allen & Unwin, 1932), pp. 77ff.

[11] For these and many other hacker quotes, see Steven Levy, *Hackers: Heroes of the computer revolution* (New York: Anchor, 1984).

ing still, for the hi-tech Faustian, is intervention into biology. One guru in this field, Marvin Minsky, describes the human mind as a 'meat machine', and argues that since meat is rather an inefficient medium for thought, our descendants should have microchips inserted into their brains: otherwise they will remain, like us, only 'dressed-up chimpanzees'. The capacity to create people as one wants them to be is, one imagines, the ultimate ambition of the Faustian will.

The thrust of a final set of claims is that, properly deployed, technology is a force for political freedom, for democracy. It has long been viewed, by some, as having at least a democratizing tendency—through rendering traditional economic privileges obsolete, or by alleviating the mass poverty which breeds dictators. In the IT age, once again, new boasts are heard. According to that most dithyrambic enthusiast for technology, Buckminster Fuller, democracy can at last be 'structurally modernized [and] mechanically implemented'[12]—for example, through lightning broadcast of the information the public needs if it is effectively to monitor government. Governments, in their turn, can, through E-mail bombardment, be quickly apprised of public opinion on the measures they are contemplating. Not a few futurologists envisage a network of computer users tired, apparently, of violent or 'erototronic' video games engaging, instead, in political debate; a hi-tech resurrection, on a grand scale, of the participatory democracy of the Athenian *agora*.

III

I now turn to the enslavement camp. What do those who speak of technology as a force against freedom have in mind? Their claims can, once more, be squeezed under three headings. First, and most familiar, is the Frankenstein thesis. In intent, technology should help us control our environment and lives, but in practice its products have risen against us and gone out of our control, individual and collective. This thesis takes several, sometimes incompatible, forms. 'The handmill gives you society with the feudal lord, the steam mill society with the industrial capitalist', wrote Marx, inspiring a doctrine, 'technological determinism', espoused by many of his followers, if not by him.[13] Human beings cannot control the basic thrust of technological development, since the con-

[12] Quoted in Ferré, *Philosophy of Technology*, p. 62.

[13] Quoted in G. A. Cohen, *Karl Marx's Theory of History* (Oxford University Press, 1978), p. 41.

trolling instruments—political and legal systems, and ideologies—are themselves inexorably shaped by the 'material forces of production'.

At some odds with this view is the fear that recent technology provides massive scope for the control of the many by the few. Buttons pressed in Washington or Moscow decisively and instantly affect the lives of millions around the globe. And it won't be 'we', but 'they', who decide the forms that the genetically or cybernetically engineered creatures of the future take, humans included. At odds both with technological determinism and the scenario of a world governed by a handful of technocrats is the worry voiced by Alasdair MacIntyre. Nearly all of us have become impotent, but not because of subjection to iron laws of production or a 'well-organized system of social control'. On the contrary, technology is responsible for a 'lack of control... at the heart of the social order'.[14] Ours is the impotence that obtains where chaos rules, where sheer scale, complexity and variety defy survey and order. I lack power, but not because someone else has it. Power, in Foucault's metaphor, is no longer located in the heart of the social body, but suffused throughout its uncountable capillaries.

A related set of claims and images, some of them in competition, is to the effect that technology endangers democratic freedom. Here, too, the older nightmare of 'Big Brother' commanding technical resources of surveillance and propaganda, beyond the dreams of a Stalin, has yielded to the sense that the power of the State, as much as that of its citizens, has been eroded. For it is technology which has spawned multi-national companies outside democratic controls; enabled speculators so to rock the markets that policies on which governments were elected are shipwrecked; and left decision-making to computers whose original programmes were shaped by goals and assumptions which no one can recall and for which no one in particular can be held accountable. Other complaints are heard. It is all very well to bombard the White House or Downing Street with E-mailed protests, but would a government which made policy on the hoof in response to such grousings and grumblings be behaving responsibly? And if Herbert Marcuse is right, that is all the protests would be, for 'technological progress ... creates forms of life which ... reconcile the forces opposing the system and ... defeat ... all [real] protest in the name of ... freedom from toil and domination'.[15] Such is the superficiality of debate in technological society that protest is reduced to a 'phone-in' from Mr Angry of Luton.

[14] *Herbert Marcuse* (New York: Viking, 1970), p. 80.
[15] Quoted by Alasdair MacIntyre, *Herbert Marcuse*, p. 70.

Marcuse's charge leads on to a final set of claims, levelled at the constricting effects of technology upon the human psyche and our conception of it—at epistemic enslavement. It is a fair bet, for example, that the creatures who are to succeed us 'dressed-up chimpanzees' will not be designed by Minsky and his colleagues for enhanced sensitivity or deeper contemplative wisdom but, to use the favourite term, for their 'smartness'—their speed and efficiency at processing information. We are already familiar, from those who speak loudest on such matters, with the equation of mind with processing abilities. That is hardly surprising, given that those whom the media and conference organizers seem to prefer for speculating about mind or consciousness are people whose philosophical psychology is inspired by analogies with computers.

If technology helps to narrow our conception of mind so, according to a related charge, does it serve to narrow, to the point of exclusion, our perspectives on the world about us. Heidegger's famous accusation against the technological 'way of revealing' the world was that it 'drives out' alternative ways of revealing it. The world then becomes so much 'equipment' or 'standing reserve' to be tapped at will. The Rhine ceases to be a waterway uniting communities, and becomes merely a source of hydro-electric power and part of the tourist industry; the pasturing cows are only tomorrow's hamburgers, not the beasts which once had their individual names and shared in the farmer's family life.[16] All of us, even the most pure aesthetes, must sometimes view the things and creatures around us in pragmatic, equipmental terms. What distinguishes the technological view, for Heidegger, is its potential for being the only one, for we *can* get through life seeing everything in nothing but those terms. People myopically impressed by the world as an object of beauty or worship die out. Those who are myopically impressed by it as a source of energy do not: they even prosper.

IV

For dramatic effect, I have referred to two camps—those of liberation and enslavement—from which there issue claims for and against technology as a force for freedom. It would be more accurate, however, to speak of alliances and misalliances among the claims, which, like military ones, are loose and shifting. Certainly it is possible to combine some liberationist claims with some enslavement ones, and we have already noted tensions within each

[16] *The Question Concerning Technology and Other Essays* (New York: Harper & Row, 1977), pp. 26ff.

camp. This is unsurprising, given that different criteria for freedom have been at work. There is no reason in principle, for example, not to combine the Promethean conception of technology as the creative exercise of will-to-power with the Frankenstein spectre of it running out of its creators' control. Such, indeed, was Spengler's position. For him, recall, technology is 'the destiny of the free personality'. But he observed that, already by the 1930s, 'the creature is rising up against its creator', that 'the lord of the world is becoming the slave of the machine'.[17] Technology, he predicted, would create intolerable traffic jams, mass unemployment, colonial revolt, and even global warming, all of which will conspire to destroy the whole enterprise. But until that day arrives, Faustian man remains the fullest expression of freedom that history has produced.

Again, it is possible to share the Marcusian sense that technology narrows our cultural and epistemic horizons with optimism about its democratizing potential. This seems to be the position of the post-modernist writer, Jean-François Lyotard. Having complained of a tendency to restrict knowledge to what can be 'translated into quantities of information', into commodities, he admits the possibility that 'the computerization of society' might also facilitate forms of political 'terror'. But Lyotard prefers to believe that it will, instead, promote a 'politics that would respect... the desire for justice'. This will occur provided that there is free public access to data banks, so that millions of citizens, ensconced by their terminals, can join in the 'language game of perfect information'.[18]

Where claims which I have sketched really do conflict, empirical resolution is sometimes invited. If true at all, it is true as a matter of fact, presumably, that technologically induced ecological changes will produce global effects beyond our control. Other conflicts, though, call for more by way of judgment and evaluation. Talk of technology smothering 'real' protest, say, or 'driving out' valid perspectives on the world, is hardly neutral, so that people who agree on the facts might disagree as to the cogency of such descriptions.

Instead of trying directly to adjudicate all, or even some, of the conflicting claims, I want to bring out a broad theme, with its various sub-themes, that has lurked beneath the preceding discussion, occasionally surfacing.

---

[17] *Man and Technics*, pp. 90ff.
[18] *The Postmodern Condition: A report on knowledge* (Manchester University Press, 1986), pp. 4 & 67.

David E. Cooper

V

I find it harder to state the general theme than the sub-themes into which it devolves, but perhaps it is this: our notion of the person or self, and its modification by technology and its ethos. Everyday notions of the person, in contrast with the rather bloodless ones that preoccupy too many philosophers, are intimately connected with those of responsibility and autonomy. Crudely, an aspect of someone's life is an aspect of him or her as a person if it belongs in the sphere for which they are responsible, within which they have governance over their life. The suggestion I shall explore is that technology and the practices and attitudes it inspires serve to reduce this sphere, and hence to narrow our understanding of the aspects of people's lives which constitute *them*, which characterize their persons or selves. This suggestion is, I suppose, akin in spirit to those obituaries—'the death of the subject', 'the death of the self'—which Parisians are fond of pronouncing. But I won't explore this kinship here.

This suggestion, if well-taken, produces an amended version of the enslavement thesis. It is not that technology has made persons less free than they were, on some abiding, traditional conception of the person. Rather, it erodes the traditional conception through shrinking our understanding of what belongs in the sphere of the person, of his or her government and accountability. If so, technology is not 'modern' only in the obvious sense: for it aids and abets a tendency which some commentators regard as distinctive of 'modernity', from the Enlightenment on—namely the reduction of the self to a 'point', abstracted from the larger fabric of life.[19]

I divide my broad theme into three sub-themes, which I label those of 'self-reliance', 'the integrity of the person', and 'one's life-as-a-whole'.

VI

Asked what he thought of Western civilization, Gandhi famously replied that it would be a good idea. One of his complaints was about modern Western medicine, for it amazed him that the average European was so ready to 'hand over custody of his body to the experts ... as if it were an appendage ... for which he bore no responsibility'.[20] Such medical practice, he saw as reinforcing a

[19] See, for example, Alasdair MacIntyre, *After Virtue* (London: Duckworth, 1982), and Jürgen Habermas, *The Philosophical Discourse of Modernity* (Cambridge, Mass.: MIT Press, 1987).

[20] Bikhu Parekh, *Gandhi's Political Philosophy* (London: Macmillan, 1989), p. 26.

14

pernicious tendency to divorce the self from the body, the latter becoming a mere possession, like a car. The point is well-taken, for many people do seem to subscribe to the assumption, generally mute, that if the experts can fix it up, it's not one's own responsibility. After all, if there's a technical fix for it, then it's a *condition*—something you find yourself in or landed with. Such may be the reasoning of those food and pin-ball 'addicts' in America who now successfully sue their employers for failing to 'respect' their condition by not installing larger chairs in the office or pin-ball machines in the canteen.

It is not only in the broad area of medicine that the sense of something's being a matter for expert judgement and fixing erodes self-reliance. One witnesses, for example, a remarkable growth, especially in the USA, in the number of people handing over the choice of a partner to computerized dating or 'nuptial facilitator' agencies. The choice of a husband, wife, or lover, for such people is no longer an act of personal expression and commitment, but the hi-tech equivalent of the arranged marriage, the outcome of a piece of information processing.

The general point illustrated by these examples is well put by Christopher Lasch in *The Culture of Narcissism*:

> The conversion of popular traditions of self-reliance into esoteric knowledge administered by experts encourages a belief that ordinary competence in almost any field, *even the act of self-government*, lies beyond the reach of the layman.[21]

Some readers may reflect, for instance, upon the influence of the current directive in British universities—fuelled surely by the increasingly technical nature of much research—that students must first be instructed in how to do research before actually doing any. Upon its influence, that is, on people's relationship to their work and its place in their lives. *Prima facie*, it is difficult for someone to regard the performance of intellectual labour as an expression of self if the rules for its performance are dictated by experts on 'research methods'. Other readers may reflect on whether people's sense of *home*, and the relation between home and personal identity, is altered in an age when, due partly to the technical savvy required, they must increasingly surrender the care and repair of their houses to professionals. Professionals, one might add, who, unlike the traditional village craftsman, are likely to be complete strangers to those employing them.

[21] *The Culture of Narcissism* (New York: Norton, 1979).

VII

Shrinkage of the fields in which people can regard themselves as self-reliant or self-governing is not the only way in which the ethos of technology reduces the sphere of the person. It does so, too, by reinforcing distinctions among the various faculties of which the self, traditionally conceived, was the locus—knowledge, reason, taste, sensibility, sensuality, emotion, moral sense, and others. A handful of these are then privileged to the point that the rest become excluded from our notion of self or personhood. The rich integrity of the self is broken up, and only a fragment remains.

The 'neo-Luddite' Theodore Roszak remarks, 'we do not bring the full resources of the self to the computer'. Before the PC screen, 'sensual contact, intuition, inarticulated commonsense' must be 'largely ... left out'.[22] His point is that modern technology, generally and not just IT, places a premium upon certain endowments and capacities which, precisely because of the centrality of technology to contemporary life, are then deemed to be the quintessential attributes of a human being. We have already noted, for instance, the narrowing of knowledge and reason to the acquisition and processing of information.

One obvious victim of this tendency is the wisdom and understanding of tradition. With its 'cult of the future... its indifference to the past and mistrust of [inherited] thought', writes Milan Kundera, our technological age 'deprives people of memory and... retools them into a nation of children'. Technology is 'organized forgetting'.[23] But with this amnesia, there is also lost the sense of oneself as, in essence, an inheritor of a form of life that stretches back: the sense, as Herder put it, that 'a man's humanity is connected by a spiritual genesis ... with his countrymen and forefathers'.[24] A distance opens up between a person and whatever traditions may still impinge on him or her. They become data, ornaments of life to pick or reject, no longer integral elements of life. It is no objection to these observations to point to today's almost obsessive rhetoric of 'respect' for traditions and 'roots'. On the contrary, such rhetoric is only needed, only possible indeed, in a climate where what is supposed to be 'respected' is already eroded. As Al-Ghazali long ago remarked, those who live within tradition

[22] *The Cult of Information*, p. 71.
[23] *The Book of Laughter and Forgetting* (Harmondsworth: Penguin, 1983), pp. 235–6.
[24] *On Social and Political Culture* (Cambridge University Press, 1969), p. 312.

do not make a noise about it, nor even self-consciously recognize
that they do so.

Victims of the technological ethos, too, are those 'resources of
the self' which, today at least, are treated as disjoint from the faculties of knowledge and reason. Within a society geared to technological progress, and honouring the faculties which contribute to
it, moral conviction, religious sense, taste and sensibility play only
bit parts. They are then relegated to the realm of the merely 'subjective', but not, ironically, with the intent that they be regarded as
fundamental aspects of the subject, the self. On the contrary, these
days one is said to 'choose' one's form of religious or moral
'expression', but without committing the one sin that remains, that
of rejecting the 'choices' made by anyone else. To those one must,
like the 'last men' whose emergence Nietzsche predicted, cheerful-
ly bray 'Yeah! Yeah!'[25] The implication, of course, is that taste,
morality and religion do not constitute a person's being: they
belong, rather, on the *smorgasbord* from which the person—essentially a creature of knowledge and reason, but regrettably unable to
live by these alone—can pick and choose at will. Such, it seems, is
the Manichean divide between the faculties essential and inessential to the technological enterprise apparent in today's educational
ideology. The main principle laid down by the Collège de France
in its report on education to the President was that in a 'well-
attuned education', there shall be a sharp distinction between 'the
universalism integral to scientific thinking' and the 'relativism'
endemic to the humanities—even to such judgments, presumably,
as that Bach was a composer of genius, or that the practice of *suttee*
is evil.

VIII

I dub my final sub-theme that of 'a life-as-a-whole'. Anything recognizable as our traditional conception of the person would, I suggest, never have emerged if the life of the average person had not
possessed a certain stability and continuity—not merely the spatio-
temporal continuity defined by having the same body from birth
to death, but the kind implied by talking of a life as possessing
'narrative structure'. A person, that is, was a creature who was not
merely self-governing in this or that field, but one who could give
predictable shape and structure to a life-as-a-whole; who, having
reached maturity, could successfully plan, within obvious limits,
for life.

[25] *Thus Spake Zarathustra*, in *The Portable Nietzsche* (New York:
Viking, 1954), 'Prologue'.

David E. Cooper

It is a sociological banality that existence in an age of rapid and ubiquitous technological change impairs any such capacity. The kind of work for which someone trained disappears, as chimney-sweeps and coalminers have discovered. Or it has remained, but anyone over forty is reckoned incapable of accommodating to the latest innovations. Or consider the mobility, the sudden change of location, which people must accept if they are to remain employed. To recall an earlier point, the incapacity to plan with confidence for a career, for where one will live, for how, if at all, one will slot into society, is due, not to subjection to the iron directives of a planning authority, but to a technological society's lack of a centre. It is the relative chaos which has replaced the framework provided by the traditional, gradually evolving norms and ways of earlier societies that disrupts the possibility of a life-as-a-whole, and with it the conception of a person as the creature whose life that is.

It is a cliché of contemporary sociology, too, that in technological society, there is an erosion of intimate human relationships. Affluence makes it possible, and the nature of work makes it sometimes imperative, for children to fly the coop at a young age. Job mobility and the range of work now open to women help contribute to increased divorce rates, to the point where few newly-weds expect, or perhaps even hope, that this will be their last marriage. No man, we are told, is an island. A life-as-a-whole is not only led in interaction with other such lives, but must be: for the stability and structure I am able to lend my own life depends, in part, on those people with whom I am most intimately engaged lending a like stability and structure to their lives. As intimacy is eroded, so then is my possibility of a life-as-a-whole.

Affluence, job mobility and the like are not the only threats to intimacy offered by technology. The hacker who spoke of being able to 'create your own universe' on the computer screen goes on to say, 'You can do anything you like. You don't [even] have to deal with people.' The nature of many people's work indeed becomes increasingly solipsistic. Even the friendly departmental secretary is a thing of the past: at any rate, you see much less of her now that you E-mail your letters instead of dictating them. Don DeLillo nicely encapsulates the contemporary phenomenon as he boards his plane: 'Air travel reminds us who we are. It's the means by which we recognize ourselves as modern. The process removes us from the world and sets us apart from each other.'[26]

---

[26] *The Names* (London: Picador, 1988), p. 254.

Do the Successes of Technology Evidence the Truth of Theories?

MICHAEL SMITHURST

Borrowing perhaps from mathematics, there is a custom of speaking of science as pure science and applied. Platonism, and other classical positions in the philosophy of mathematics, did not think of the applications of mathematics as a test of the truth of its theorems.[1] But the picture is otherwise for science and technology. It is initially tempting to say that the theories of pure science are empirical generalizations and that the applications of these theories in the makings and doings of technology, accordingly as they succeed or fail, test the theories. Qualifying factors and counter-acting causalities needing to be allowed for, falsification will not be immediate, but inexplicable and apparently irremediable technological failure is likely to be taken as falsifying a theory, and a continued and expanding pattern of technological achievement, a triumph of technology, as the superannuated trope has it, will be taken as a confirmation of a theory, from the inductivist perspective, adding in spadesful to the evidence for its truth.

That such a view of the relation of science to technology has been held is clear from an argument that has familiarly been made against Karl Popper's account of scientific method. G. J. Warnock and Hugh Mellor, amongst others, have put the following argument against Popper.[2] Popper claims to have solved Hume's problem of induction, and to have solved it by showing that there is no such thing as induction, either as a logical principle, or as a mode of reasoning, or as a method of discovery. The problem of the inductive leap, from a finite set of singular propositions as premises to a universal conclusion, is solved, says Popper, when we understand that there is no leap to be made. In reply to this, Warnock and Mellor[3] argue that Popper's critical method simply

[1] But emphatically these are not the only positions. See C. Ormell (ed.), *New Thinking about the Nature of Mathematics* (Norwich: MAG-EDU University of East Anglia, 1992).

[2] G. J. Warnock, review of *The Logic of Scientific Discovery* by Karl R. Popper, *Mind* LXIX (1960), 99–101, D. H. Mellor, 'The Popper phenomenon', *Philosophy* 52 (1977), 195–202.

[3] Mellor 'The Popper phenomenon' puts the point dramatically: 'A century of electromagnetic theory has transformed radio from the merest speculation to the firmest of facts ... the whole point is that we have more

transfers the onus of making the inductive leap from those who invent theories to those who make use of them. It may be that the work of the pure scientist can be plausibly represented in terms of conjecturing and refuting, but the fact of induction is not thereby disproved, but evaded, for the inference to the character of future cases from facts about past cases, that Hume takes as a given, is made by whoever designs a steam engine or builds a bridge.

A different but related figure which has been used to explicate the relation of pure to applied science is employed in the following analogy. Technology may be conceived as embodying the actions of science. Theories are beliefs and technological endeavours the actions in which those beliefs are made manifest. It is a familiar thesis that beliefs are not simply causes of actions but so related to them that belief attributions are, at least in some cases, logically inseparable from propensities to action. 'The character of the belief in the uniformity of nature can perhaps be seen most clearly in the case in which we fear what we expect. Nothing could induce me to put my hand into a flame–although after all it is *only in the past* that I have burnt myself ... The belief that fire will burn me is of the same kind as the fear that it will burn me', says Wittgenstein (*Philosophical Investigations*, 472, 473).[4] Hume is only able to get his problem of induction off the ground by assuming an artificial distinction of 'agent' and 'philosopher'. My practice you say refutes my doubts. But you mistake the purpose of my question. As an agent, I am quite satisfied in the matter; but as a philosopher who has some share of curiosity, I will not say scepticism, I want to learn the foundation of this inference' (*Enquiries*).[5] But in other places this a distinction that Hume, the naturalist, when assessing the authenticity of beliefs, will not countenance:

[4] L. Wittgenstein, *Philosophical Investigations* (Oxford: Basil Blackwell, 1958). The references are to numbered sections.

[5] D. Hume, *Enquiries Concerning Human Understanding and Concerning the Principles of Morals*, P. H. Nidditch (ed.) (Oxford University Press, 1978), p. 38.

reason to expect Mr Pye's transmitters to work than Mr Heath Robinson's. Popper is indeed hot for rationality; but by divorcing it from reasons for anticipating one future experience rather than another, he deprives it of much of its import. Obviously we have such reasons, and the progress of science has given us many more, as everyone's actions every day attest that they believe. Why will Popperians not admit to such beliefs, which they reveal every time they turn on the light or use the telephone?

> The Roman Catholics are certainly the most zealous of any sect in the christian world; and yet you'll find few among the more sensible people of that communion, who do not blame the Gunpowder-treason, and the massacre of St Bartholomew, as cruel and barbarous, tho' projected or executed against those very people, whom without any scruple they condemn to eternal and infinite punishments. All that we can say in excuse for this inconsistency is, that they really do not believe what they affirm concerning a future state; nor is there any better proof of it than the very inconsistency.[6]

These images, theory and evidence, belief and action, may seem reasonable as first stabs at elucidating the notions of pure and applied, but a look at the history of technology soon diminishes their plausibility.

Consider, say, the invention of printing. Gutenberg or Coster may be credited with it, and what they probably did was first devise printing with moveable type. Undoubtedly a technology, appearing and then flourishing, is to be seen here. About 40,000 books are estimated to have been in print within fifteen years or so of Gutenberg's 42-line Bible of 1454. What this development looks like is the putting together of antecedently existing craft skills. Blocks, moveable type, printer's ink (carbon particles suspended in linseed oil instead of water), paper, presses and ratchets, seem orchestrated by inventive ingenuity, yet Fust printed in multicoloured inks in 1400, the *Diamond Sutra* of 868 was a printed scroll, printed paper money circulated in Szechuan in the ninth century, and so on. Not only does this not look like actions in which theory is made manifest, it does not even look as if any phenomena are being produced which theory is needed to explain.

Perhaps this is not a good example, so consider instead the history of the external combustion engine. Forget the Greeks. Pre-existing theory seems to get a grip on Thomas Savery's device patented in 1698, the 'Miner's Friend', a steam engine for raising water from flooded mines. The operative principle is that the cooling of a steam-containing vessel condenses the steam to a few drops of water thereby leaving a virtual vacuum. Water can be raised through valves into the evacuated vessel. Newcomen's improved engine of 1705 had a piston in a cylinder and used air pressure to push down the piston. Both machines depended on cooling the chamber after each heating, and were slow and dangerous. It was not until 1782 that Watt made a steam engine that got

[6] D. Hume, *A Treatise of Human Nature* (Oxford University Press, 1978), p. 115.

three times as much work out of a ton of coal as Newcomen's. Watt's central improvement was to keep the steam chamber hot and lead the steam into a separate condensing chamber to cool. Thereafter, efficiency continually increased through the use of hotter steam at ever higher pressures. I say that theory informs Savery's engine because the invention would not seem possible without the seventeenth-century debate on the reality of the vacuum and the experiment of Torricelli and Viviani in 1644. Air as a distinct substance having a density and exerting pressure long had a loose metaphysical familiarity, but Boyle's work gave it calculable consequences eventually of moment in steam engine design.

Yet most explanatory theory came not before but after the development of steam engines. Steam blows out of the picture when in 1824 Sadi Carnot reasons that the efficiency of an idealized engine depends only on the temperature of its hottest and coldest parts and not on the substance driving the mechanism. His biographer describes this graduate of the Ecole Polytechnique as driven to conceive a theory of heat engines because 'offended that the British had progressed so far through the genius of a few engineers who lacked formal scientific education'. Carnot's work on the ideal cyclical sequence of changes of pressures and temperatures in a fluid activating an engine, was incorporated in the 1850s into the work of Clausius and Kelvin, and is effectively the beginning of thermodynamics. So mostly, in this story, theory follows practice.

In seeking an example of theory purely conceived then successfully applied, so vindicating the theory, the history of electrical technology may look more promising. 'It's blue, it crackles, and it comes in zig-zags', said Franklin, when asked why he hypothesized that a fluid flowing in storm clouds was identical with one elicited when he combed his hair. In so far as electromagnetic phenomena are natural, their manifestations are individual and dispersed. Theory is required to unite them. In the sixteenth century Gilbert had theorized that the earth was a magnet, and seen the attractive and repulsive powers of lodestones as kin to those found in fricated amber (*elektron*) and glass. In the early eighteenth century, Desaguliers investigated the properties of what he named 'conductors' and 'insulators'; and in 1745 von Kleist contained the elusive fluid in 'Leyden jars'. Capacitors had been invented. In the 1750s, technology, of a sort, throbs, as lightening conductors sprout across Europe and North America. It is impossible to know whether or not to call what comes next 'technology'. Devices are invented, but for the laboratory, for the purposes of theory, and to feed the inquisitiveness of discovery. In 1785 Coulomb, an engineer, following Priestley, begins the quantification of electrical science, and shows that the attraction or repulsion between charges varies inversely as the square of the distance. In 1800 Volta makes an electric cell from which current can be drawn off continuously. It is then used by Humphrey Davy, to pull apart the atoms of tightly bound molecules. Preparations are made of the metals sodium, potassium, magnesium, calcium, strontium and barium. Faraday, his assistant, was later to state the laws of such molecule-breaking electrolysis. If supporting evidence for something is being produced here, it is as much for the atomic theory as for generalizations about the nature of electricity. Practical applications beyond the theoretical follow within a generation. William Sturgeon produced an electromagnet in 1823, and by 1831 the American Joseph Henry had made a version that could lift a ton of metal. Faraday's dynamo was invented in the 1830s, and versions of it were used to power machinery in the next decade, but it was not until 1872 that von Hefner-Alteneck patented the first really efficient generator, enabling electricity to be produced in quantity from combustible fuels and running water. Theory ran alongside these developments. In the 1860s Maxwell mathematicises Faraday's work on lines of force and produces a unitary mathematical treatment of electricity and magnetism. It is not until the end of the century that theoreticians understand how chemical reactions involve electron transfers and interpret electric currents in terms of the flow of electrons. Yet the theory requisite had been in place since Davy's time, when, in 1878 and 1879, Edison made hundreds of trial and error experiments to produce a light bulb, finding a successful filament, at last, in a scorched cotton thread. Theory intertwines with technical endeavour in this history, but arguably, better than the previous two examples, it fits a model under which a discovery is made, laws are formulated and tested, then entrenched by an expanding technology.

Twentieth-century physics and chemistry are the best places to look for instances of technology that are primarily theory-driven. The atom bomb sprang fully armed from the head of theory. A particle accelerator does not, as far as I know, do anything but accelerate particles. But the construction and operation of such a machine is a move in pure science. Confirmation of theory is its *raison d'être*. It is technical, but not technological.

I have tried to show that it is not all that true that technology is the use of theories, but in so far as it is the use of theories, is it right to say that technological successes inductively confirm the theories used?

There is an obvious objection to such a picture of how successful technology is related to theoretical science. This is that the connections between a theory and a technical device may be more or less loose. Duhem taught us long ago that theories are not confirmed or infirmed in isolation. What a host faces what Quine called the tribunal of experience when a developed piece of technology is put into operation. The empiric generalizations of materials science, and the homely saws of paperclip improvisation, jostle in the dock with portentous truth from half a dozen branches of science. Where the verdict is 'innocent', all must be accruing confirmatory evidence at once. The fact is that often, but not always, in an *experimentum crucis*, the apparatus used is unproblematic for either side. Technology is commonly used to get evidence for theories, but the general relation of technology to theories seems too slack to be characterized as evidence.

One good reason for saying that applied science does not furnish evidences for pure science is this. Theory, large theory, paradigmatic theory, is defeasible, but technology is not. Theory can get dislodged, but technology and its successes always stay. Scientific fallibilism does not imply that theory is revisable in any direction. We can move from the belief that the earth is round to the belief that it is an oblate spheroid, but not that it is flat. Whatever we think in future about electromagnetism, however electromagnetic phenomena are re-conceptualized in the light of deeper theory, television sets and washing machines will not be dismissed as illusions. The ubiquitous phenomena of electrical technology now belong with the inductive certainties of common life that any theory must be consistent with.

The concept of inductive certainty, of empirical certainty effectively, is a contested one, but it has perhaps unexpected friends. David Hume is one. Near the beginning of the essay on miracles, he distinguishes those judgements of experience that amount to proofs from those that are just probabilities. 'Why is it more than probable', Hume asks, 'that all men must die; that lead cannot, of itself, remain suspended in the air...?', and so on.[7]

Twentieth-century logical empiricism represented 'empirical' propositions as merely probable. So popular was the idea that nothing synthetic is certain, that both logical positivism and its opponents presented versions of it. One root of the idea was phenomenalism, under which a singular empirical proposition may accumulate, but never complete, a sequence of confirmatory observations alleged expressive of its 'content'. That being the case for singular propositions, it seemed axiomatic that something as

---

[7] Hume, *Enquiries Concerning Human Understanding*, p. 114.

presumptuous as a generalization from experience could never be 'conclusively verified', and hence lacked certainty. Karl Popper, disavowing this epistemology, nevertheless had his own version of the thesis. Singular propositions, of necessity expressed in general terms, have nomological implications. The unverifiability of laws thus communicates itself to singular propositions, and entails that no theory for which some singular proposition stands as a confuting instance is ever, in fact, conclusively falsified.

But analytical philosophy of the earlier twentieth century has also another strand, equally empiricist in origins and sympathies, that pulls against the logical empiricist denial of certainty. Moore and Malcolm, in varying styles, berated impossibly demanding criteria of knowledge, and affirmed against 'the philosophers' the certainties of common language and common sense. Austin ironically enquired should he chew on a telephone to assure himself of its reality, and hearing a pig, and smelling a pig, and then seeing it, knew there was a pig, no doubt about it, 'the question is settled'.[8]

Not in the mainstream of this debate, but one of its most interesting extractions, is Wittgenstein's exploration of its themes in *On Certainty*; interesting partly because it strikingly anticipates the tension between justificatory epistemology and naturalism so characteristic of the analytical epistemology of the later twentieth century.

Wittgenstein says, in a striking phrase, that the fact that water boils and does not freeze in such and such circumstances 'is fused into the foundations of our language game'.[9] In *On Certainty* he returns constantly to the thought that 'not everything that has the form of an empirical proposition is one' (ibid. 308). He says that 'the truth of certain empirical propositions belongs to our frame of reference' (ibid. 83). Despite the fact that technology sometimes comes out of theory, the truths and practices of technology are thought of as constituting part of the background against which testing takes place, part of the system of unquestioned assumptions that make testing possible at all. That has not always been, and may not always be, their place; for

> the same proposition may get treated at one time as something to test by experience, at another as a rule of testing (Ibid. 98) ... All testing, all confirmation and disconfirmation of a hypothesis takes place within a system. And this system is not a more or

[8] J. L. Austin, *Sense and Sensibilia* (Oxford University Press, 1967), pp. 115 and 123.

[9] L. Wittgenstein, *On Certainty* (Oxford: Basil Blackwell, 1979), 558. References to *On Certainty* are to numbered sections.

less arbitrary and doubtful point of departure for all our arguments: no, it belongs to the essence of what we call an argument. The system is not so much the point of departure, as the element in which arguments have their life. (ibid. 105)

Seeking alternatives to the foundationalist metaphor as deployed in classical empiricist epistemology, Wittgenstein uses the figure of a riverbed:

It might be imagined that some propositions, of the form of empirical propositions, were hardened and functioned as channels for such empirical propositions as were not hardened but fluid: and that this relation altered with time, in that fluid propositions hardened, and the hard ones became fluid ... And the bank of the river consists partly of hard rock, subject to no alteration or only to an imperceptible one, partly of sand, which now in one place now in another gets washed away or deposited. (ibid. 96,99)

The indefeasibility of technology locates its truths and practices as part of the riverbed.

Suppose, madly, a freezing of all technology under a green despotism. What would follow? Technology generates data which even now flood the maw of theory. Witness the Hubble telescope and its abundance of information that NASA hopes one day, time and money permitting, to sift and use. Under this true conservatism, innovations of method would be forbidden, but data obtained by existing methods could still be collected. This will resolve some disputes, but eventually new methods will be required for determination of theory choice. And until some choices are determined, we will not know which roads to travel. So theorizing may slowly come to a halt. Alternatively, irresoluble theory choices may accumulate, as theoretical imagination multiplies hypotheses, all divorced from practice, and all compatible with, and the expression that uncomfortably suggests itself is, the facts. And this is where the fantasy crumbles. One conclusion to take from Wittgenstein is that there are certainties, but no class of uninterpreted intrinsically evidential facts. Practical investigation and theorizing are so intertwined that one cannot make sense of an injunction to suspend one and continue the other. Even the most theoretical of physicists can never usurp the philosopher's armchair.

Our understanding of theories and evidence has singular origins. It hiccups its way out of the epistemologies of scepticism and the refutation thereof. Hume's problem must be resolved, positivism swallows the toad whole, Popper elaborates a solution that expels the very notion of confirming evidence, subsequent demonstrations that this falsifies the history and sociology of science flirt with relativism, whilst roundly declaring scepticism to be 'not interesting'. Perhaps the best way out of this is to recognise that science is a common enterprise and theories public possessions.

An American scholar who had completed a book on Darwinian evolutionary theory published an article in a British newspaper in which he expressed his dismay at the frequency with which his fellow-countrymen volunteered their disbelief of Darwin's theory.' I am a trained engineer', said one, 'and I can assure you that God created the universe about ten thousand years ago.' Bishop Ussher would have disagreed, of course. But, in fact, the engineer's is no longer a tenable opinion amongst others. The integumental binding of science makes it impossible. You cannot keep physiology and genetics, and dispense with evolution. You cannot have chemistry and mechanics, and leave behind astronomy and geology. Technology interweaves the parts of science and theories rest on theories. Whatever the virtues or vices of reductionism, it is this consideration that delivers the unity of science.

If a conflagration destroyed all books and records, and, of the human race, left just half a dozen English-speakers alive, would the English language still exist? A working vocabulary of 40,000 words would, but the descendant evolved in a couple of generations would soon be less penetrable than, say, the English of Malory. Science, like language, is not primarily located in any one head. From a holocaust-reduced society of six, even if some of them were scientists, precious little would get through to the next generation. You may be able accurately to describe a theory, say the theory of evolution (to take an easy one), and you may even know a lot of the arcane detail. Perhaps you can specify some of its more striking predictive successes, say, that of the dual mandibular structures consequent on the evolution of the auditory ossicles.[10] But without the records, without the aid of the anatomist, and without the technology of palaeontology, there is not much chance that this bit of the theory, or much else of it, can be re-established as scientific knowledge. The link of belief with action is broken. It is not so much that technology evidences theories, as sets the conditions under which they can truly be scientific theories at all. In the absence of technology, our verbal characterizations of theories, fished from memory, would soon become

[10] For an account of this, see S. J. Gould, 'An earful of jaw', *Eight Little Piggies, Reflections in Natural History* (London: Penguin Books, 1994), pp. 95–108.

metaphysical myths, like the atomic theories of the Greeks. If technology vanishes, so, for the most part, does pure science; for it is technological detail that locks theories onto experience, and, without it, we have only a story, an imagination of things, as insubstantial as the eschatological beliefs of Hume's sensible Roman Catholics.[11]

[11] I am grateful to Paul Gilbert and Nick May for discussion of topics in this paper.

Instrument and Reality: The Case of Terrestrial Magnetism and the Northern Lights (Aurora Borealis)

WILLEM HACKMANN

In recent years there has been an increasing focus on the role of instruments in the study of nature, both by historians and by philosophers of science,[1] and even by a few art historians who are interested by the images produced by these devices.[2]

My own approach is that of the historian of science with an

[1] See in particular David Gooding, Trevor Pinch and Simon Schaffer (eds), *The Uses of Experiment. Studies in the Natural Sciences* (Cambridge University Press, 1989), especially the Introduction which clearly sets out the aims of this book. My own first attempt on this subject was 'The relationship between concept and instrument design in eighteenth-century experimental science', *Annals of Science*, vol. 36 (1979), pp. 205–224, primarily based on my work on eighteenth-century electrical instrumentation and experiments, and which caused one philosopher to comment that I had used the word 'interaction' in at least five different ways.

[2] Two recent books are Martin Kemp, *The Science of Art* (London: Yale University Press, 1990) and John Gage, *Colour and Culture* (London: Thames and Hudson, 1993). Kemp's title is misleading as he deals primarily with colour theory and perspective, as he points out in the sub-title, 'Optical Themes in Western Art from Brunelleschi to Seurat'. Indeed perspective is the most common 'scientific' theme dealt with by art historians, see for instance the fine study by S. Y. Edgerton, *The Heritage of Giotto's Geometry. Art and Science on the Eve of the Scientific Revolution* (Cornell University Press, 1991). Other aspects that could be studied with profit are (a) how scientists produced idealized images of their experiments often with the help of artists, (b) how such images were used to communicate their results, (c) how artists used scientific metaphors in their paintings (of which Hans Holbein's 'The Ambassadors' (1533) is a well-known example), and (d) how artists used scientific aids (such as Dürer's perspective frame, and later the camera obscura and camera lucida) in their compositions. On (a) and (b) see the papers in R. G. Mazzolini (ed.), *Non-Verbal Communication in Science Prior to 1900* (Florence: Leo S. Olski, 1993), in particular my paper 'Natural philosophy textbook illustrations 1600-1800', pp. 169–196, and John J. Roche, 'The semantics of graphics in mathematical natural philosophy', pp. 197–233, which have useful bibliographies.

interest in the philosophical implications.[3] To the instrument historian, science is applied technology, put to the task of understanding nature by revealing the kinds of beings of which she is ultimately composed.[4]

Scientific instruments have become indispensable in collecting and 'dissecting' natural phenomena from the seventeenth century when the techniques were developed that are at the roots of modern western technology-orientated science.[5] At the most general level, these instruments made visible that which could not be seen by the unaided senses: Galileo Galilei's surface of the moon as seen through his primitive telescope, or Robert Hooke's compound eye of the fly as seen through his microscope, and in more recent times, C. T. R. Wilson's particle tracks in the cloud chamber.[6]

None of the images produced by these devices was, of course, strictly neutral, in the sense that in their making a complex relationship existed between the observer and the observed. In order to enable fellow microscopists to obtain the *same* image of the compound eye of the fly, Hooke described exactly in his

[3] My study of these implications is mainly restricted to experimental nature philosophy, the precursor of physics, although some of my observations concerning analogy and model experiments can be applied more generally, see for instance, David Gooding *et al.*, *The Uses of Experiment*, the 'Standards and models' section of the Introduction, pp. 3–4. Most studies by instrument historians have not been philosophical but artifact-orientated or in terms of cultural artifacts, see my discussion in 'Instrumentation in the theory and practice of science: scientific instruments as evidence and as an aid to discovery', *Annali dell'Istituto e Museo di Storia della Scienza di Firenze*, vol. 10, pt 2 (1985), pp. 87–115.

[4] Here I am in agreement with Rom Harré, 'The dependence of "hi-tec" science on technology through the rôle of experimental technique', chapter X in W. D. Hackmann and A. J. Turner (eds), *Learning, Language and Invention* (Aldershot: Variorum, 1994).

[5] Not only began nature to be seen in terms of a giant machine, but machines were developed to elucidate and simulate the phenomena in the laboratory.

[6] See Peter Galison and Alexi Assmus, 'Artificial clouds, real particles', in Gooding *et al.*, *The Uses of Experiment*, pp. 225–274. They demonstrate that Wilson's cloud chamber only makes historical sense when studied in the context of the confluence of two traditions at the Cavendish Laboratory: the analytic and mimetic, the analytical research in the structure of basic matter (J. J. Thomson's 'transcendental physics') and the reproduction of natural phenomena (in Wilson's case, weather phenomena such as cloud formation).

Micrographia (1665)[7] where to place the microscope in relation to the illumination of the specimen—any other arrangement would not achieve the 'true' image.[8] This begs, of course, the question of what was the true image:[9] a question not considered by Hooke in any deep sense because of his empirical instrumental approach to natural philosophy.[10] His image of the compound eye has proved

[7] I have used the 1667 edition of R. Hooke, *Micrographia: or Some Physiological Descriptions of Minute Bodies Made by Magnifying Glasses* (London, 1667), Preface, and pp. 175–180, Pl. XXIV. Catherine Wilson, 'Visual surface and visual symbol: The microscope and the occult in early science', *Journal of the History of Ideas*, vol. 69 (1988), pp. 85–108, has rightly called this image a 'landmark in scientific iconography'. See also J. T. Harwood, 'Rhetoric and graphics in *Micrographia*' in M. Hunter and S. Schaffer (eds), *Robert Hooke New Studies* (Bury St Edmunds: Boydell Press, 1989), pp. 119–147, which illustrates the subtle changes between Hooke's manuscript drawings and the published engravings, and Svetlana Alpers, *The Art of Describing. Dutch Art in the Seventeenth Century* (John Murray in association with Chicago University Press, 1983), pp. 72–118, esp. pp. 73–74, 83–84 (Hooke and microscopy). Alpers works has fascinated historians of science without gaining universal approval from art historians.

[8] Thus instrumentalists had to devise a kind of grammar to make communication possible on how their instruments had to interact with nature. For a key paper, see M. A. Dennis, 'Graphic understanding: instruments and interpretation in Robert Hooke's *Micrographia*', *Science in Context*, vol. 3 (1989), pp. 309–364, which has not yet received the serious response which it deserves.

[9] Indeed Gaston Bachelard has argued that the introduction of the microscope was an impediment: 'In truth it was only a case of spinning out the old dreams with the new images which the microscope delivered. That people sustained such excitement over these images for so long and in such literary form is the best proof that they dreamed with them', *La Formation de l'esprit scientifique*, 4th ed. (Paris, 1965), p. 160.

[10] Hooke discusses his instrumental approach in 'A General Scheme, Or Idea of the Present State of Natural Philosophy, and How its Defects may be Remedied by a Methodical Proceeding in the Making Experiments and collecting Observations Whereby to Compile a Natural History, as the Solid Basis for the Superstructure of True Philosophy', published by R. Waller (ed.), *The Posthumous Works of Robert Hooke* (London, 1705), pp. 1–75, especially pp. 35–37. For a brief discussion and further references, see my paper 'Attitudes to natural philosophy instruments at the time of Halley and Newton', *Polhem*, vol. 6 (1988), pp. 143–158, esp. pp. 143, 146–149. Hooke was well aware of the power of his pictures.

to be a success in that it has been replicated by the most modern of technologies.[11]

It could be argued that what makes an instrument scientific (that is successful in scientific terms) is not the device per se but the success of its manipulator. A powerful example is Anthony van Leeuwenhoek's simple lens microscope with which he made more scientific discoveries in the 1680s than all the contemporary 'superior' compound-lens microscopes put together.[12]

No scientific apparatus is intrinsically self-evidently superior. Its value lies in its power of persuasion. Bachelard has likened an instrument to un théorème réifié, and sociologists have analysed the part played by instruments in the scientific and psychological strategies developed to reach a consensus of opinion about a particular theory.[13] From a historian's point of view such analyses are often ahistorical,[14] but at least they have resulted in a focus on one of the key issues exercising the minds of a number of instrument

[11] J. Burgess, M. Marten and R. Taylor, *Under the Microscope. A Hidden World Revealed* (Cambridge University Press, 1990), pp. 55 and 61, figs 3.22 and 3.33. Figs 1.1–1.4 are a good example of the microscope as a technical/observational frontier: unexpected features are revealed as magnification increases, in this case the magnified head of a pin reveals the rod-shaped bacteria. On the visual impact of these technology-driven images, see John Darius, *Beyond Vision* (Oxford University Press, 1984), but which does not discuss philosophical implications.

[12] Success in these terms is judged (a) by the discovery of a new phenomenon, in Leeuwenhoek's case that of bacteria, and (b) persuading fellow practitioners that this phenomenon really exists, primarily by making replication possible (see note 8). In the case of Faraday, the only way he could get fellow practitioners to replicate a specific experiment was by sending them a miniature version of the apparatus.

[13] See G. Bachelard, *Les intuitions atomistiques* (Paris: Presses Universitaires de France, 1933), *La formation de l'ésprit scientifique* (Paris: Presses Universitaires de France, 1938), and *L'activité rationaliste de physique contemporaine* (Paris: Presses Universitaires de France, 1951), and for a discussion of his work, see S. Gaukroger, 'Bachelard and the problem of epistomological analysis', *Studies in the History and Philosophy of Science*, vol. 7 (1976), pp. 189–244, and S. Schaffer, 'Natural philosophy', in G. S. Rousseau and R. Porter (eds), *The Ferment of Knowledge. Studies in the Historiography of Eighteenth-Century Science* (Cambridge University Press, 1980), pp. 77–91.

[14] A classic example is T. S. Kuhn's account of the discovery of the Leyden jar in *The Structure of Scientific Revolutions* (University of Chicago Press, 1962), pp. 61–62, which Kuhn agrues as 'theory-induced'. I would describe such a development much more in terms of organic development and natural selection. See also note 47.

historians: What is the role of instruments in bringing about changes in scientific beliefs?

Before I proceed I had better define the different categories of scientific instruments and then focus on the category most pertinent to this paper. Some instruments, such as armillary spheres and orreries were models of how the natural world was perceived. Others, such as clocks, chemical balances, electrometers, galvanometers, and graduated astronomical angle-measuring instruments, were tools of measurement. Increasing their precision made scientific breakthroughs possible but, of course, not inevitable. Tycho Brahe's new angular measurements were necessary for Kepler's work, and Flamsteed's for Newton. A third category, such as telescopes and microscopes, could be classed as observational instruments with which nature was observed passively, while a fourth category, such as electrical machines and air pumps, allowed the manipulator to become an active participant in nature's laboratory, in this case in the space between the electrodes, or in the artificial space inside the bell jar. Thus, certain instruments could be passive or active explorers of nature. This can be a fruitful distinction when analysing the development of the scientific method. However when looked at historically, these categories blur at the edges, in the same way that armillary spheres when suitably arranged could be used for astronomical observations.

Anthony van Leeuwenhoek's microscope was the most scientific because of the way it was used (which led to its scientific success). In the same way, instruments such as Newton's prism, can be used passively or actively,[15] although it can be argued that there is generally a class of difference between 'optical instruments' such as telescopes and microscopes, and 'philosophical instruments', such as air pumps and electrical machines. Telescopes and microscopes did not 'rearrange' nature, but revealed hitherto unsuspected phenomena and structures not observable with the naked eye. The distinction between passive (purely observational) and active (phenomena-interactive) instruments may help us to understand the basic features of experimental philosophy as reflected, for example, in the study of atmospheric electricity,[16] or of the aurora boralis.

[15] Newton turned an optical toy that could be bought at any fair into a scientific instruments, see Simon Schaffer, 'Glass works: Newton's prisms and the uses of experiment', in Gooding *et al.*, *The Uses of Experiment*, pp. 67–104, esp. p. 78.

[16] I have made a similar study of this topic in 'Instruments and experiments. The case of atmospheric electricity in eighteenth-century Holland' *Tijdschrift voor de Geschiedenis der Geneeskunde,*

Willem Hackmann

The focus of this paper is on what were commonly called the 'philosophical instruments', developed for natural philosophy, which evolved into physics during the nineteenth century. From the Scientific Movement onwards, these technological devices have been used in the two basic types of experiments: (a) to discover the properties of specific natural phenomena such as the electric spark, or the vacuum, and (b) model experiments.[17]

In 1600 'electricity' referred only to the property of attraction of the rubbed amber, but more and more properties were added as these were discovered by increasingly sophisticated electric devices: repulsion (1620), glow (1705), transmission or conduction (1720), spark and shock/pain (1730), heat (1740), so that by the

[17] For my most detailed discussion on this topic see my 'Scientific instruments: models of brass and aids to discovery', in Gooding *et al.*, *The Uses of Experiment*, pp. 31–65.

Natuurwetenschappen, Wiskunde en Techniek, vol. 10 (1987), pp. 190 (60)- 207 (77). See also my essay review, 'Lightning rods and model experiments: Franklin's science comes of age', *Studies in History and Philosophy of Science*, vol. 22 (1991), pp. 679–684. My distinction of passive and active instruments has not won universal acceptance, see for instance J. A. Bennett, 'A viol of water or a wedge of glass', in Gooding *et al.*, *The Uses of Experiment*, pp. 105–114, in which he demonstrates that an instrument can be transformed from being active to passive depending on its use with which I would agree. Bennett appreciates the problems of definition within the context of the various sciences (e.g. optical versus natural philosophy), but suggests that historians should preserve the distinctions (or classifications) established by the practitioners themselves. Thus, the categories 'mathematical, optical and philosophical instruments' grew out of the intellectual and craft distinctions of the late seventeenth century. However, as demonstrated by his paper 'The mechanics' philosophy and the mechanical philosophy', *History of Science*, vol. 24 (1986), pp. 1–28, Bennett in his criticism is too biased towards the seventeenth century, and even here he sticks too rigidly to what he considers to be the category of 'practical mathematics'. In fact, categories have always been quite fluid, as can be discovered when analysing the shift of the categories of instruments over time in contemporary encyclopediae, such as the various editions of John Harris's *Lexicon technicum; or an Universal Dictionary of Arts and Sciences, explaining not only the terms of Art, but the Arts themselves* (1704). Such a study has not yet been undertaken. At any time, the act of categorizing must be to some extent ahistorical. What Bennett has not appreciated is that my distinction of 'passive' and 'active' was simply to facilitate a discussion about the role of instruments in experimental philosophy—for which our language is still inadequate. Historians like Bennett have concentrated on the history of astronomy and not on the development of laboratory practices.

1740s, electricity had to save *all* these properties to be true electricity. These discoveries were technologically driven. When in the 1760s a fish was discovered that gave electric-type shocks, it was considered to possess a special type of electricity—'animal electricity'—as it did not appear to have all the properties of 'common electricity' as discovered in the laboratory. In 1776 Henry Cavendish demonstrated in a particularly elegant experiment with his model fish that the apparent dissimilarity was caused by the differences in the relationship between the various electrical factors such as intensity, capacity and resistance in the laboratory and in the fish.[18]

This brings us to the second group of experiments—the model experiment. The intuitive feeling about the underlying regularity of natural processes became the basis for two research strategies: (a) the use of the analogous argument in the framing of physical theories, and (b) the model experiment in which natural processes were 'recreated' in the laboratory by means of small models, when the strict inductive procedures could not be applied, or when no direct experimental intervention was possible. Galileo Galilei had to employ the analogous argument and assume that the patterns of light and darkness flitting over the moon's surface as seen through his telescope had the same cause as the similar phenomena he saw in the Tuscan mountains. In the same way the behaviour (and causes) of real lightning, earthquakes and other natural phenomena, including the aurora borealis could only be studied in the labo-

[18] H. Cavendish, 'An account of some attempts to imitate the effects of the torpedo by electricity', *Phil. Trans. R. Soc.*, vol. 66 (1776), pp. 196–225. Cavendish reached this conclusion by rearranging well-known electrical laboratory apparatus such as charged Leyden jars, brass chains, and electroscopes, see Hackmann. 'The relationship between concept and instrument design', pp. 221–222 and 'Instruments and experiments', pp. 55–56. Newly discovered electrical phenomena were only considered electrical if they passed what had become the traditional tests, such as charging a Leyden jar and deflecting a gold-leaf electroscope. For another case study see the identification of electricity produced by the friction of steam from steam engines by Lord Armstrong and Faraday in my paper 'Electricity from steam: Armstrong's hydroelectric machine in the 1840s', in R. G. W. Anderson, J. A. Bennett and W. F. Ryan, (eds), *Making Instruments Count. Essays on Historical Scientific Instruments Presented to Gerard L'Estrange Turner* (Aldeshot: Variorum, 1993), pp. 147–173. What is interesting in this example is that once it had been established that the prime cause of this unusual phenomenon was the friction of the water/steam particles, this was readily accepted as it did not challenge the existing 'mind set' concerning the behaviour of electricity.

Willem Hackmann

ratory by means of models. The development of such models in a specific topic as the aurora borealis illustrates the intricate relationship that evolved between theory, scientific observations and laboratory experiments.[19] It also highlights the strengths and weaknesses of experimental philosophy.

The Aurora Borealis

The aurora borealis has intrigued man since antiquity. Aristotle refers to it in his *Meteorologia* as burning flames, 'torches', and 'goats', and thought that it had the same cause as 'shooting stars'.[20] Grouping together the aurora borealis, comets and shooting stars persisted well into the sixteenth century. Scientific study began with Pierre Gassendi (1592–1655), who in 1621 gave the phenomenon its modern name ('aurora borealis'), while Giovanni Domenico Cassini (1625–1712) at the end of the seventeenth century employed the name 'northern lights', as it was most prominent in northern regions, near the earth's pole.[21] In the eighteenth

[19] For a more detailed discussion on analogies and model experiments, see Hackmann 'Instruments and experiments', pp. 45–48, and the references cited therein. Galison and Assmus (in Gooding *et al.*, *The Uses of Experiment*, p. 227) refer to the Victorian tradition of 'mimetic experimentation', and (p. 231) that a number of late nineteenth-century morphological scientists began to use the laboratory to reproduce natural occurrences such as cyclones or glaciers, etc. In 1892 the geologist E. Reyer wrote that researchers had given up either because quantitative experiments seemed impossible or because experiments had been unable to imitate (*nachbilden*) natural conditions. Now, by reproducing these phenomena, at least partially, much could be learned, thus heralding the beginning of new experimental physical geology. As I have shown, model-making was one of the main techniques of seventeenth- and eighteenth-century experimental philosophy, and in my study of the Dutch natural philosopher Martinus van Marum (1750–1837), I called it 'imitative experiments', based on van Marum's Dutch term *nabootsen*, see Hackmann, 'The relationship between concept and instrument design', pp. 220–223, based on my unpublished Belfast MA thesis (1970), 'The electrical researches of Martinus van Marum (1750–1837)', pp. 225–226.

[20] S. Gunther, 'Das Polarlicht im Altertum', *Beiträge z. Geophysisik*, vol. 6 (1904), pp. 98–107. For a good synopsis of these sources, see E. Newton Harvey *A History of Luminescence From the Earliest Times Until 1990* (Philadelphia: The American Philosophical Society, 1957).

[21] Icelandic settlers in Greenland are said to have used the same term *noroljós* for this phenomenon. For a typical late nineteenth-century textbook account, see A. Guillemin, *Electricity and Magnetism*, revised and edited by Silvanus P. Thompson (London: Macmillan, 1891), pp. 90–134.

century auroras were studied by a great many observers who described in detail the many different shapes and physical characteristics.[22]

Chemical and Optical Models

The hypotheses developed to explain auroras can be broadly categorized into three groups: magnetic, electrical, and others. Petrus van Musschenbroeeck (1692–1761) and Louis-Guillaume Le Monnier (1717–1799) suggested that they were formed from clouds of matter that, at certain times, had evaporated to the upper regions of the atmosphere, where these clouds had taken fire, becoming luminous and phosphorescent. Jean Jacques d'Ortous de Mairan's (1678–1771) popular theory that auroras (as zodiacal light) were produced by the penetration of the atmosphere of the sun into the atmosphere of the earth, was opposed by Leonhard Euler (1707–1783) who suggested that they were caused by the solar rays pushing to a great distance the particles of the upper regions of the atmosphere and then making them luminous by the reflection of the same rays on their surface. Late eighteenth-century chemists such as Richard Kirwan (1733–1812) and Antoine Libes (1752–1832), advanced the gas theory of the formation of auroras, identified as hydrogen by Kirwan and nitric acid vapours by Libes, formed by electrical discharges through nitrogen and oxygen, when illuminated by the sun. Libes was led to this theory after observing the play of reddish colours in a flask of nitric acid in sunlight in 1790. The ideas of these chemists were in line with the latest discoveries in pneumatic chemistry.[23] In the early nineteenth century, the dominant view of those not countenancing the magnetic or electrical nature of auroras, was that proposed by J. H. Vallerius in 1708, that the aurora was a reflection of the sun below the horizon.[24]

[22] For the most comprehensive history on the aurora in relation to the history of electroluminescence, see Harvey, *A History of Luminescence* (Philadelphia: pp. 37–40, 44–45, 83–84, 255–263, 399.

[23] As the study of gasses was called in the eighteenth century because they were collected over pneumatic troughs.

[24] René Descartes had already formulated a similar theory that the aurora was the result of the light of the sun below the horizon. Before the experiments proposed in the seventeenth and eighteenth centuries, the dominant view was that this phenomenon was caused by vapours from the earth set on fire.

Willem Hackmann

Electrical Models

Both the electrical and magnetic theories of the formation of auroras were heavily influenced by analogous observations and model experiments. Two key model experiments which lay at the heart of the beginning of the modern study of electricity and magnetism were William Gilbert's (1544–1603) magnetic terrella (1600) to which Otto von Guericke (1602–1686) responded with his electric terrella (c. 1660). Gilbert fashioned a model earth from a small

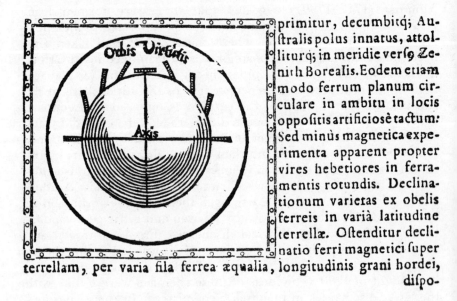

primitur, decumbitq; Auftralis polus innatus, attolliturq; in meridie verfo Zenith Borealis. Eodem etiam modo ferrum planum circulare in ambitu in locis oppofitis artificiosè tactum: Sed minùs magnetica experimenta apparent propter vires hebetiores in ferramentis rotundis. Declinationum varietas ex obelis ferreis in variâ latitudine terrellæ. Oftenditur declinatio ferri magnetici fuper terrellam, per varia fila ferrea æqualia, longitudinis grani hordei, difpo-

Figure 1. William Gilbert's 'orbis virtutis' or sphere of power about the magnetic terrella, showing magnetic dip (Gilbert, *De magnete*, 2nd ed., (1628), p.78).

spherical loadstone, with which he was able to show that magnetic dip of the mariner's compass varied with latitude.[25] It strengthened his argument that the compass was not attracted to the pole star, but that it was an intrinsic property of the earth, and that gravitation was magnetic. Von Guericke countered this with his model earth made of sulphur (as, according to him, the earth consisted

[25] An interesting aspect here is the importance of contemporary technology as the work was first undertaken by the compass maker Robert Norman in his *The Newe Attractive* (1581) in his attempt to develop a new navigational instrument.

mainly of 'sulphureous particles'), with which he demonstrated that gravitation was not magnetic but electrical, caused by the friction of the atmosphere on the rotating earth. The sulphur ball was rotated and the friction was produced by the hand rubbing on the ball's surface.[26]

This developed into the eighteenth-century theory that the earth was surrounded by an all-pervading electric fluid extending into the upper reaches of the atmosphere. Natural phenomena, such as lightning, auroras, whirlwinds, earthquakes, hail, and rain storms were caused by local disturbances in atmospheric electricity. Since during this period only electrostatic phenomena were known in the laboratory, these natural 'electrical' phenomena were also seen in terms of the behaviour of static charges. Testing this theory resulted in the development of increasingly sensitive electrometers and in a number of model experiments. Francis Hauksbee (1666–1713) in 1705 was the first to identify the electrical nature of the glow he produced in evacuated glass tubes, and this eventually led to the 'aurora flask', a pear-shaped glass vessel with a central brass electrode, partly coated on the outer surface with tin foil. It had a brass fitting to enable it to be exhausted by a vacuum pump. This popular device was used from the early 1770s

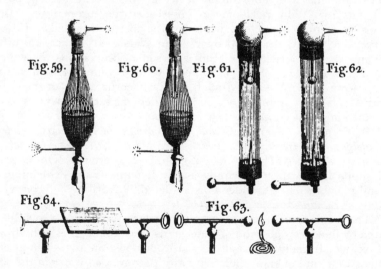

Figure 2. Aurora flasks and tubes (G. Adams, An Essay on Electricity (London, 1787), Pl. IV, figs 59–62.

[26] For fuller description and references, see Hackmann, 'Attitudes to natural philosophy instruments', pp. 152–153, Pls 3 and 4.

to model the aurora.[27] The evacuated flask demonstrated that the glow produced at the brass spike behaved differently for positive and negative charges. Similar phenomena were demonstrated by means of exhausted glass cylinders terminating inside at one end in a brass point-electrode and at the other in a ball-electrode. The most influential proposer of this theory was Benjamin Franklin (1706–1790), who regarded the aurora as an example of the electric fire which in this case was diffuse rather than concentrated as in a lightning flash.[28]

By the end of the eighteenth century the purely electrostatic theory of the aurora had declined in favour of the magnetic one, although a purely electrical model made a brief return when Gaston Plant (1834–1889) in the 1830s, utilizing the recently developed powerful electrochemical battery, suggested that auroras were produced by a flow of positive electricity—'for the luminous phenomena were the same as those of the positive electrode in the voltameter'.[29]

[27] J. Ferguson, *An Introduction to Electricity* (London, 1770), pp. 19–20, and Experiment XVII, pp. 63-65, Pl. I, fig. 4. Abb Jean-Antoine Nollet in *Leons de physique expérimentale* (Paris, 1753) had already described a similar demonstration device in vol. 1 lettre IV, pp. 80–82, Pl. I, figs 3–5. William Henley in 1774 called the aurora tube a 'glass exhausted prime conductor', made by the instrument maker Edward Nairne, but already described by William Watson in 'an account of the phenomena of electricity in vacuo with some observations thereupon', *Phil. Trans. R. Soc.*, vol. 47 (1752), pp. 362–376, who remarked that the phenomenon 'resembled very much the lively coruscations of the aurora borealis'. These immensely popular demonstration devices were described in all the contemporary textbooks.

[28] B. Franklin, *Experiments and Observations on Electricity, made at Philadelphia in America. To which are added Letters and Papers on Philosophical Subjects* (London, 1769), pp. 30–52, letter V, 'Observations and suppositions toward forming a new hypothesis for explaining the several phenomena of thunder-Gusts' (dated 29 April 1749); Harvey, *A History of Luminescence*, pp. 259–260.

[29] The 'voltameter' was a technique developed to measure quantity of charge by the decomposition of water in its constituent gases of hydrogen and oxygen. Initially it was also a test to demonstrate that current electricity was a kind of static electricity as it decomposed water in the same way. For the early history of this device (and references), see Willem Hackmann 'Leopoldo Nobili and the beginnings of galvanometry', in G. Tarozzi (ed.), *Leopoldi Nobili e la cultura scientifica del suo tempo* (Instituto per i beni artistici culturali naturali della Regione Emilia-Romagna, 1985). Gaston Plant constructed the first practical lead-acid storage battery or accumulator in 1854.

Magnetic and Electromagnetic Models

Edmund Halley (c. 1656–1743) called attention to the relationship between the auroral streamers and the earth's magnetic field in 1716. Magnetic fluid existed between the earth's crust and a solid kernel, and it was this outpouring from crevices in particular in the polar regions where the earth's crust was thinnest, that produced the luminous phenomena.[30] During the eighteenth century magnetic terrella were a popular acquisition for the amateur naturalist. Well known was Christopher Wren's (1632–1723) large magnetic sphere at the Royal Society, surrounded by thirty-two compasses for studying terrestrial magnetism,[31] and Lord Abercorn's powerful three-inch sphere in a green shagreen case. Another collector had his six-inch terrella engraved with a global map and meridional lines.[32] In the early nineteenth century the loadstone terrella was replaced by terrestrial globes with small internal magnets that 'mimicked' the behaviour of the earth's magnetic field.[33] Elaborate mechanical models were made incorporating rotating terrellae, and sophisticated models were devised to demonstrate contemporary theory concerning the causes of magnetic variation. John Canton in 1759 modelled the sun and the earth with boiling water and bar magnets to illustrate his theory that the diurnal variation was governed by solar heat.[34] This was only one of many similar such models.

The purely magnetic theory of auroras declined after the discov-

[30] According to Halley, the rotation of the internal kernal was the cause of the diurnal and annual variations of magnetic declination. A layered earth is, of course a very old idea, described for instance by Dante, and in the nineteenth century in Jules Verne's *A Journey to the Centre of the Earth* (English trans. 1874) in which the intrepid travellers discover the inhabitants the existence of which had already been suggested by Halley.

[31] W. H. Quarrell and M. Mare (eds), *London in 1710 from the Travels of Zacharias Conrad von Uffenbach* (London: Faber and Faber, 1934), p. 99; J. A. Bennett, *The Mathematical Science of Christopher Wren* (Cambridge University Press, 1982), pp. 44–54.

[32] Royal Society Journal Book, vol. 12, pp. 203–204 (15 February 1721). I am grateful to Patricia Fara for this information in her unpublished Cambridge Ph.D. thesis, 'Magnetic England in the eighteenth century' (September 1993), pp. 51 and 63.

[33] For an example made between 1791 and 1797, J. R. Edge, 'a terrestrial globe by William Bardin, with some unusual magnetic features', *Bul. of the Scientific Instrument Society*, no. 39 (1993), pp. 16–18.

[34] J. Canton, 'An attempt to account for the regular diurnal variation of the horizontal magnetic needle; and also for its irregular variation at the time of an aurora boralis', *Phil. Trans. R. Soc.*, vol. 51 (1759), pp. 400–402.

ery of electromagnetism by Hans Christian Oersted in 1820,[35] and models were now devised to demonstrate the possible electromagnetic nature of the earth's magnetic field. A persuasive impetus for the field theory of magnetism were the iron filing patterns around spherical loadstones and bar magnets, and Ampère and others found the same patterns associated with electromagnets and with wires carrying a current.[36]

The first electromagnetic terrella was made by Peter Barlow (1776–1862) in 1824, shortly thereafter followed by William Sturgeon (1783–1850), who devised many pieces of electromagnetic demonstration apparatus. One of the finest pieces was made after the design of Leopoldo Nobili of Florence. His model earth consisted of a wooden globe with copper wire wound in grooves latitudinally around the globe, supported on a brass pedestal with hardwood base. The globe can be rotated by hand and tilted at different angles. When it was connected to a battery by means of the terminal at the base, a magnetic field was formed which could be detected by a magnetic dip needle on a separate support. Thus these devices exhibited the same magnetic properties as did Gilbert's terrella more than two hundred years previously.[37]

[35] He made this discovery at a time when there was no common agreement on the physical relationship between electricity and magnetism. Oersted's conviction concerning the unity of natural forces which made him do this experiment was based not on contemporary experimental philosophy, but on the ideas of Johann Wilhelm Ritter and Kant, and the German *naturphilosophen*. See my paper 'Leopoldo Nobili', pp. 203–233, esp. pp. 210–211.

[36] Once iron filings were used to make visible the unseen magnetic fields around lodestones and artificial magnets, the representations became increasingly abstract, eventually being simply curved lines in space, see Hackmann, 'Scientific instruments', p. 192, but especially David Gooding, 'Magnetic curves and magnetic field: experimentation and representation in the history of a theory', in Gooding *et al.*, *The Uses of Experiment*, pp. 183–223, esp. pp. 186–190. For a more detailed discussion see David Gooding, *Experiments and the Making of Meaning* (Dordrecht, Boston, London; Kluwer, 1990), pp. 95–113.

[37] Barlow made his electromagnetic terrella at the instigation of George Birkbeck of the London Institution to illustrate experimentally that Ampère's hypothesis that all magnetism was due to electric current could explain the terrestrial magnetic field. W. Sturgeon, 'Account of an improved electromagnetic apparatus', *Annals of Philosophy*, vol. 12 (1826), pp. 357–361. Sturgeon's electromagnetic globe is reproduced in Gooding, *Experiments and the Making of Meaning*, pp. 206–211, figs 7.12 and 7.13. Nobili's globe is described in his *Memorie ed osservazioni edite ed inedite del Cavaliere Leopoldo Nobili colla descrizione ed analisi de' suoi apparati ed instrumenti*, 2 vols (Florence, 1834), vol. 2, pp. 22–24, Pl. IV, fig. 3,

Figure 3. Leopoldo Nobili's electromagnetic terrella *c*. 1830. Magnetic dip would be demonstrated by a small magnetic needle suspended above the globe (Nobili, *Memorie ed osservazioni*, pp. 22–24. Pl. IV, fig. 3).

described by Hackmann and P. Brenni, 'Gli strumenti scientific', in the exhibition catalogue *L'eredità scientifica di Leopoldo Nobili* (Comune di Reggio Emeila: Biblioteco Municipale 'A Panizzi' and Instituto e Museo di Storia della Scienza, 1984), item 26, pp. 73–74, and in my *Catalogue of the Pneumatic, Magnetic, Electrostatic, and Electromagnetic Instruments in the Museo di Storia della Scienza* (Florence: Giunti, 1995), item 249.

Willem Hackmann

A fruitful model was proposed by Faraday based on his work on magnetic induction. He considered the earth as a great magnet rotating, and thereby giving rise to currents which flowed out of its poles into the upper air, thence returning at the equitorial regions. Auroras were caused by discharges which constituted a part of these perpetual currents, but he never constructed a laboratory model to demonstrate his ideas.[38]

In the 1840s a new source of high-voltage electricity was being developed—the induction coil.[39] This allowed for detailed studies of the behaviour of such discharges in the high vacuums of a new generation of vacuum discharge tubes, such as those developed by Johann Heinrich Wilhelm Geissler (1815–1879).

Powerful electromagnets demonstrated the deflection of the voltaic arc (the new powerful electric sparks) by the magnetic poles, and the same was demonstrated to be the case of luminous discharges in vacuum tubes. This led in 1849 to a popular demonstration by Arthur-Auguste de la Rive's (1801–1873) device in which the luminous isocharge slowly rotated around the soft iron core of an electromagnet placed inside a spherical vacuum vessel. In essence this was a modification of the eighteenth-century aurora

[38] Michael Faraday, *Experimental Researches in Electricity*, vol. 1 (London, 1839), article 192. He visualized the terrestrial globe, surrounded by its magnetic lines of force, revolving on its own axis, and as it cut through its own lines of force generating electricity, the currents moving from the equator through the earth to the poles, and there leaping into the air to return to the equator through space, possibly in the form of the aurora borealis and australis. This is an interesting example of the transference of ideas, in this case from magnetic induction in the laboratory to the earth as a huge electromagnetic generator similar in action to the electromagnetic machine being developed from Faraday's work. Faraday was interested in the aurora phenomenon for quite some time judging by his annotations dated from 1832 to 1849. See L. Pearce William, *Michael Faraday* (London: Chapman and Hall, 1965), pp. 207–208, 224, note 11. See also Bern Dibner, *Faraday Discloses Electro-magnetic Induction* (Burndy Library, 1949).

[39] On its early history, see W. D. Hackmann, 'The induction coil in medicine and physics 1835–1877', in C. Blondel, F. Parot, A. J. Turner and M. Williams (eds), *Studies in the History of Scientific Instruments* (London: Roger Turner Books for the Centre de Recherche en Histoire des Sciences et des Techniques de la Cité des Sciences et de l'Industrie, Paris, 1989), pp. 235–250.

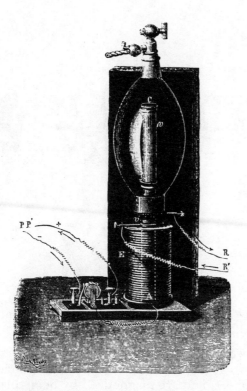

Figure 4. De la Rive's (1849) rotating luminous electric discharge around the pole of a magnet (Guillemin, *Electricity and Magnetism*, p. 449, fig. 304).

egg in which the electrodes have been replaced by the electromagnet.[40]

These experiments indicated a connection between magnetism or electromagnetism and the luminous discharge, which led de la Rive to devise the most sophisticated electromagnetic terrella in about 1859, by which he demonstrated the relationship between terrestrial magnetism and the aurora borealis. A wooden sphere

[40] A-A. de la Rive, 'Extrait d'une lettre à M. Regnault (sur les Aurores boréales) [regarding effect of magnet on electric discharge in gas and the aurora borealis]', *Comptes rendus Ac. Sci.*, vol. 29 (1849), pp. 412–415; also in *Phil. Mag.*, vol. 35, pp. 446–449. See also the following other papers by de la Rive: 'On the diurnal variation of the magnetic needle, and on the aurorae boreales' *Phil. Mag.*, vol. 34 (1849), pp. 286–294, in which he demonstrates that electric sparks are effected by the poles of a strong electromagnet from which he moved on to his aurora experiment: 'I cannot conclude this abstract without drawing attention to the circumstance, that M. Arago had already pointed out in 1820, shortly after Œrsted's discovery, the possibility of acting upon the voltaic arc by this magnet, and the analogy which might result between this phenomenon

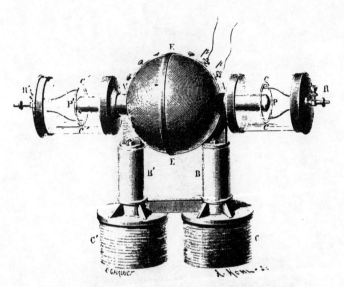

Figure 5. De la Rive's (1859) polar auroras model (Guillemin, *Electricity and Magnetism*, p. 121, fig. 69).

representing the earth has at its extremities two stems of soft iron which rests on two vertical cylinders of soft iron which also serve to support the sphere, and are magnetized by being placed on the poles of a powerful electromagnet. The two stems projecting from the sides of the globe are enveloped in an evacuated glass cylinder. To work the model, the globe is surrounded by two strong bands of blotting paper (one around the equator, the other running from pole to pole), and wetted with salt water. When an induction coil is connected to the equitorial band running to the two soft iron projections, discharges occur inside the glass envelopes, sometimes in one and at other times in the other, but rarely in both at the same time. As soon as the soft iron is magnetized by the electromagnet, these luminous jets spread out forming round the soft iron stems 'animated with the movement of rotation, the direction of which depends on that of the magnet-

and that of the aurora borealis' (p. 294); 'On the rotation of the electric light round the Pole of an electro-magnet', *Phil. Mag.*, vol. 15 (1858), pp. 463–466, which refers to his description of this apparatus in his textbook translated by C. V. Walker, *A Treatise on Electricity in Theory and Practice*, 3 vols (London, 1853–1858), vol. 2 (1856), p. 308; 'On the action of magnetism upon the electric discharge in highly rarified gaseous media', *Phil. Mag.*, vol. 33 (1867), pp. 512–530: 'I demonstrated the existence of this action as early as 1849, by showing that a magnetic pole causes jets of electricity which escape from it radically to rotate'. See also Harvey, *A History of Luminescence*, p. 29.

ism'. Thus, de la Rive concluded that his model reproduced the behaviour of the aurura.[41] Work on the aurora continued unabated but our case study ends here.[42]

Discussion and Conclusions

In an attempt to define the relationship between available technology and the domain of phenomena, Rom Harré has formulated three domains of beings relative to the possibility of human observation. The first, the 'domain of actual existence', consists of phenomena that can be observed and identified without the help of sense-extending instruments. The second contains beings which could exist (they are plausible) but the sense-extending instruments to reveal them do not (yet) exist. The third domain contain beings which are beyond all possible experience. The ontological basis of science is not only fixed by the choice of source models for theories, but also by the available technology. The invention of new instruments will cause the boundaries between these domains

[41] Guillemin, *Electricity and Magnetism*, pp. 119–122, fig. 49. I know of three versions of this rare apparatus, one in Museo Scientifico dell'Instituto di Fisica G. Marconi dell'Università in Rome, signed 'N. 496 Societé genevoise/ 113 Plainpalais/ Genève', another in the Conservatoire des Art et Métier in Paris, while the prototype is in the Musée d'Histoire des Sciences in Geneva, made in the workshop of Professor Thury in about 1859 under the direction of Eugène Schward, a skilled German instrument maker. It is described by de la Rive in 'Nouvelles recherches sur les aurores boréales et australes, et description d'un apparail qui les reproduit...', *Mémoires de la Société de Physique et d'Histoire naturelle de Geneva*, vol. 16 (1862), pp. 313–342 and in 'Description d'un appareil qui reproduit les aurores boréales et australes avec les phénomènes qui les accompagnent', *Comptes rendus Ac. Sci.*, vol. 54 (1862), pp. 1171–1175, and in English 'Further researches on the aurorae boreales and the phenomena which attend them', *Phil. Mag.*, vol. 23 (1862), pp. 546–553.

[42] It is now thought that the primary cause of the aurora is charged particles which approach the earth at high speed and are deflected by the earth's magnetic field into a ring. These aurora-producing particles may well be part of the outer Van Allen belt. Spectroscopic studies have indicated that the auroral light is radiation from atoms and molecules of oxygen and nitrogen, with some radiation by hydrogen atoms. Auroral displays are most frequent during heightened sunspot activity in particular when there is a solar flare burst out. Magnetic storms occur at the same time. The simultaneous activity of sunspots and magnetic storms was already observed in the eighteenth century, but the significance not understood until the twentieth.

to change, although the development of the instruments them-selves may well be independent of theories, such as the manufac-ture of better optical glass. Thus, Rom Harré argues that there is a dialectical relationship between science and technology as our capacity to use cognitive skills requires an input from material skills.[43]

This case study bears out Rom Harré's analysis, although my focus has been more on the interaction between observation and modelling. Direct observations with instruments and the analo-gous argument illustrated by conceptual models yielded two basi-cally different types of experiment. The first consisted of labora-tory procedures isolating a specific phenomenon (e.g. the electric spark) and determining its properties (e.g. the spark is hot); the second of laboratory models with which to imitate the natural phenomenon as perceived (e.g. the aurora is magnetic). Each of these strategies led to the development of experimental appara-tus.

The inspiration for these processes was by no means just scien-tific in the narrow modern sense, but also cultural. Von Guericke's electric terrella was inspired not by research in electricity, but by his views on cosmology and Robert Boyle's (1627–1691) work on the vacuum not by physics but by theology. Similarly, Joseph Needham has argued that the earliest Chinese magnetic compass was a lodestone modelled on the Great Bear, and made to rotate on the 'heaven-plate' of the diviner's board, thus it was the product of contemporary Chinese philosophical thought.[44]

The process of determining the properties of a phenomenon was usually technology driven, but within an overarching theoretical world view or conceptual framework, which influenced not only what the observer saw but also how he interpreted his observation. During most of our aurora case study, the overarching theory con-cerned the behaviour of matter—initially whether natural phe-nomena were due to an emanation or exhalation of a subtle sub-stance, and (post-Gilbertian) whether they were caused by the mechanical behaviour of corpuscles. Imbedded within this frame-work was conceptualizing a phenomenon (or family of phenomena) in terms of a more restricted framework, which in the case of the

[43] Rom Harré, 'The dependence of "hi-tec" science'.

[44] On the Ham diviners and the lodestone spoon, see J. Needham, *Science and Civilisation in China*, vol. IV: 1 (Cambridge University Press, 1962), pp. 261–269, and for details of the cultural transmission of the magnetic compass, W. D. Hackmann, 'Jan van der Straet (Stradanus) and the origins of the mariner's compass', in Hackmann and Turner, *Learning, Language and Invention*, chapter IX.

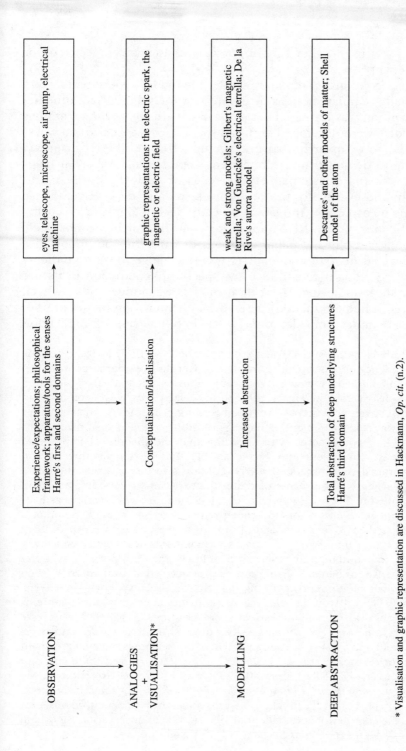

Figure 6. Visualising and modelling nature with instruments.

* Visualisation and graphic representation are discussed in Hackmann, *Op. cit.* (n.2).

aurora was in terms of contemporary gas chemistry, electricity or magnetism.[45]

The second group of experiments, the model experiments, were especially popular in areas of nature in which direct experimental intervention was limited, as was the case with the aurora. Families of models were developed within a specific conceptual framework, in the case of magnetic auroral models, from Gilbert's magnetic terrella to demonstrate (mimic) the properties of terrestrial magnetism, to the electromagnetic terrellas of Barlow, Sturgeon and Nobili, and eventually to the aurora terrella of de la Rive. These models incorporated the latest magnetic technology and discoveries. Obviously de la Rive appreciated that the earth was not made of wood, wet blotting paper and coils of wire, but the importance for him and his fellow modellers was to arrange those ingredients that were thought to produce the specific phenomenon in nature under study. The success of their models was judged by how well they replicated (simulated) the natural phenomenon (or set of phenomena) under consideration. Indeed the success of replication

---

[45] Both Hacking and Gooding have argued for the possibility of the independence of experiment from theory. Certainly exploratory work can create phenomena for which no contemporary theoretical explanation exists—discoveries in eighteenth-century electricity is a good case in point, but even here there were broad guiding principles, such as whether nature was corpuscular and behaving according to the laws of mechanics. Although I would agree with Gooding that the theoretical hypotheses guiding an experiment may fail to specify the relevant parameters and observational conditions necessary to obtain a result, and that these had to be learned in the process of doing the experiment. Gooding refers in this context to Trevor Pinch who has identified such observations as having a *low externality*—they are not predicted by, or dependent upon a particular theory. referring again to my own work on eighteenth-century electricity, I cannot think of any experiment that was totally conceived outside any contemporary theoretical framework. I am not sure whether the same would be true of the images produced by the microscope, such as Hooke's compound eye of the fly or Leeuwenhoek's bacteria. However, here I would claim that these images are still within a contemporary cultural framework, based on assumptions to which Mari Hesse refers as *coherence conditions*. See Gooding, 'Magnetic curves and magnetic field' pp. 191–192, and his 'How do scientists reach agreement about novel observations?', *Studies in the History and Philosophy of Science*, vol. 17 (1986), pp. 205–230; M. B. Hesse, *Revolutions and Reconstructions in the Philosophy of Science* (Hassocks: Harvester Press 1980), pp. 131–134; T. Pinch, 'Towards an analysis of scientific observations: the externality and evidential significance of observation reports in physics', *Social Studies of Science*, vol. 15 (1985), pp. 3–36.

was their sole criteria since there is no logical means of testing their validity.[46] There could be only weak or strong models depending on how closely they were tied to the observed phenomenon.[47] As is demonstrated by the history of the aurora borealis, specific models generally went into decline not because they became less successful in reproducing the phenomenon within their own framework, but because this conceptual framework was superseded by a more successful one, for example, the decline of the eighteenth-century gas model in favour of the nineteenth-century electromagnetic model.[48] In all this technical development played an essential role.

Perhaps I should end with an intriguing paradox highlighted by the history of scientific instruments. This shows that the more sophisticated the apparatus became, the further the observer moved away from nature, his senses becoming subservient to increasingly complex technological tools. Thus, in our western technology-based science the observer's experience of natural phenomena has become more and more indirect. On the one hand instruments have been wonderful facilitators, on the other they have become increasingly daunting psychological barriers between us and nature.

[46] A common strategy adopted in the eighteenth century by an opponent of a particular model was to argue that the modeller was involved in a circular argument. According to this, the same concepts used to make the model where then 'proofed' by the model if it behaved as predicted or recreated the phenomenon. It is difficult to see how analogies can fully avoid circular arguments.

[47] The same is true of the analogous arguments: Galileo's comparison of the surface of the moon with the mountains of Tuscany is more directly related to the observed phenomena than Hooke's extension to this analogy in which he suggests that a more powerful microscope would perhaps show that the moon is populated by sheep grazing on grass like that covering the hills of Salisbury Plain—see Waller, *The Posthumous Works of Robert Hooke*, p. 243. In the case of model experiments, their success depends on how close the parameters of the real phenomenon can be modelled or paralleled in the laboratory. Cavendish's model of the electric fish was far more successful than Priestley's electric earthquake model in which he passed an electric discharge between two wet planks (the earth) the force of which toppled blocks of wood (the buildings) (Hackmann, 'Instruments and experiments', p. 52). This paper has only dealt with the empiricism of modelling, not with such important practicalities as scaling.

[48] To reiterate note 14 above, instead of Kuhn's paradigm shift, it seems to me that a great deal of this development can be seen in organic or evolutionary terms, with natural selection taking place between competing models.

Realism and Progress: Why Scientists should be Realists

ROBIN FINDLAY HENDRY

Introduction

For as long as realists and instrumentalists have disagreed, partisans of both sides have pointed in argument to the actions and sayings of scientists. Realists in particular have often drawn comfort from the *literal* understanding given even to very theoretical propositions by many of those who are paid to deploy them. The scientists' realism, according to the realist, is not an idle commitment: a literal understanding of past and present theories and concepts underwrites their employment in the construction of *new* theories. The theme of this book is philosophy and technology, and here's the connection: *new* theories point out—and explain—*new* phenomena. So realism, claim the realists, is at the heart of science's achievement of what Bacon, that early philosopher of technology, identified as science's aim: *new* knowledge offering *new* powers.

How does this become an argument for realism? Scientific realism enters the story twice: (tacitly) adopted by *scientists*, it motivates scientific practice, while the success of the practice might support realism as a *philosophical* view of science. To fill the story in, we need to know what the realist view is, and an account of how a scientist who accepts it would behave differently from one who does not. But there are many ways to be a realist about science, of which the following is a representative sample:[1] (Aims) Science aims to provide literally true stories about the world, to be a process of *discovery* and of *explanation* rather than one of *construction* and of *saving the phenomena*. (Semantics) Theoretical discourse is to be construed *literally*: uses of theoretical terms are putative references to theory-independent entities—whether observable or not. Literally construed, theoretical claims can be true or false independently of our interests and commitments. (Epistemology) Given the success of our best theories, we have reason to believe they are (approximately) *true* and referentially successful, not just empirically adequate. The different compo-

[1] Here I (roughly) follow B. C. van Fraassen, *The Scientific Image* (Oxford University Press, 1980), ch. 2. The list is not meant to be independent, exhaustive or acceptable to all realists: I have tried to select *methodologically relevant* claims.

nents work together: in accepting a theory, we take it that it advances our scientific aims, so acceptance of a theory must involve the (tentative and qualified) belief that it is *true*. Since realist aims are more stringent, the realist's deeper commitment reflects *stricter* criteria of adequacy. If these extra criteria concern *explanation*, the realist's deeper commitment results from (what amounts to) an inference to the best explanation. Given that we *do* have predictively successful, explanatory theories, we have reasons to think them (approximately) *true*.

With an eye on the three components of realism, we can see that some feature of scientific practice might commit practitioners to realism if it is appropriate only to realist aims, reveals realist semantics, or would not make sense unless some theory was taken to be *approximately true* on account of its predictive success and explanatory power. Now let *methodological realism* be the claim that some practices that are central to the success of science reveal realist commitments in any of these ways. Arguments for scientific realism from methodological realism have been presented in two forms: the *explanationist* argument due to Richard Boyd and Hilary Putnam, and the vindicationist version presented by R. I. G. Hughes. The explanationist version has it that only realism itself can explain the success of the realist practices. The vindicationist argument eschews explanation: some realistic commitments are *presupposed* by scientific practice, and where the practice is successful, both practice and presupposition are *justified*. In what follows I will critically examine these arguments, raising the following issues: (i) whether—and in what way—features of scientific theorizing[2] commit scientists to realism; (ii) whether these features contribute to success; (iii) whether a convincing argument for realism can be grounded in positive answers to (i) and (ii). I will conclude *yes* to (i) and (ii), but *no* to (iii): the anti-realist can accept—and understand—the implicit recommendation in methodological realism.

1. The Explanationist Argument

In his middle period, Putnam often adduced scientific practice in his arguments against positivist philosophy of science. In

[2] Thus I will leave aside the causal arguments advanced by I. Hacking (*Representing and Intervening*, Cambridge University Press, 1983), N. Cartwright *How the Laws of Physics Lie*, Clarendon Press, Oxford, 1983) and R. N. Giere (*Explaining Science*, University of Chicago Press, 1988) that (speaking of) manipulation in experimental contexts commits us to more than mere theorizing does.

'Explanation and reference', for instance, he charged that positivist analyses of meaning fail to do justice to scientific *usage* of either theoretical terms or the theories in which they are embedded. To understand science, claimed Putnam, we need realist semantics and a realist account of theory acceptance.

To take one of Putnam's examples,[3] it is common to conjoin accepted theories and look to their joint consequences for novel predictions. The realist—for whom acceptance means *acceptance-as-true*—finds this easy to rationalise: if T_1 and T_2 are true, then so must be T_1 & T_2. For the non-realist, however, acceptance means only *acceptance-as-empirically-adequate*, and it is less than obvious that the empirical adequacy of T_1 and T_2 separately must imply the empirical adequacy of T_1 & T_2: indeed, T_1 and T_2 might be *inconsistent*, and therefore *trivially* empirically inadequate. So the pooling of explanatory and predictive power—central to the very cumulativity of science—seems inexplicable on any non-realist account of acceptance. Putnam's other examples continue in similar vein: in a good theory, scientists seek the realist's *theoretical plausibility* rather than the positivist's *simplicity*; auxiliary hypotheses are not the *minor premises* in *deductions of observational consequences*, but rather *further facts* to be filled in, so the aim of research must be *fact-finding* rather than *theory-testing*.

So where are these examples leading? In 'Explanation and reference', Putnam is content to show that the positivists failed to do justice to scientific practice, but in 'The meaning of "meaning"', there is the outline of a *positive* argument for realism:

> It is beyond question that scientists use terms as if the associated criteria were not *necessary and sufficient conditions*, but rather *approximately* correct characterizations of some world of theory-independent entities, and that they talk as if later theories in a mature science were, in general, *better* descriptions of the same entities that earlier theories referred to. In my opinion the hypothesis that this is *right* is the only hypothesis that can account for the communicability of scientific results, the closure of acceptable theories under first-order logic, and many other features of the scientific method.[4]

[3] See H. Putnam, 'Explanation and reference' in G. Pearce and P. Maynard (eds), *Conceptual Change* (Dordrecht: Reidel, 1973), pp. 199–221. References are to the reprint in Putnam's *Mind, Language and Reality* (Cambridge University Press, 1975).

[4] H. Putnam, 'The Meaning of "Meaning"', in K. Gunderson (ed.), *Language, Mind and Knowledge: Minnesota Studies in the Philosophy of Science* 7, (Minneapolis: University of Minnesota Press, 1975), pp. 131–193, at p. 155.

Robin Findlay Hendry

The force of the argument has to be in the *centrality* of the realist features of practice to the success of science, whether in communication or prediction: otherwise the realist practices—and the semantics—might reflect only so much decorative embroidery on the tapestry of science. In *Meaning and the Moral Sciences*,[5] Putnam fills out the argument (but famously re-interprets it), avowedly following Boyd in the details.

Boyd's defence of realism is set within a naturalistic epistemology that provides a detailed mechanism by which realist methods contribute to the success of science.[6] First define the *instrumental reliability* of a method in terms of the instrumental reliability of theories:

> Call a theory instrumentally reliable if it makes approximately true predictions about observable phenomena. Call a methodology instrumentally reliable if it is a reliable guide to the acceptance of theories which are themselves instrumentally reliable.[7]

Boyd's explanationist strategy is as follows: (i) Identify a reliable methodological principle or strategy of theory construction, and show that the principle can be rationalized only given realist inferences and a realist construal of theories. (ii) Show that the employment of that principle 'contributes to the likelihood that accepted theories will be good predictors of the behaviour of observables').[8] (iii) Claim that realism provides the only plausible explanation for the *reliability* of that principle.[9] Boyd's various examples of realist methods—united under a theme of the unity of science—arise in all areas of scientific practice: construction of theories, design of experiments, and assessments of the degree to which a given body of evidence supports a theory. For example:

[5] H. Putnam, *Meaning and the Moral Sciences* (London: Routledge & Kegan Paul, 1978).

[6] See R. Boyd, 'Realism, underdetermination and a causal theory of evidence', *Nous* 7 (1973), pp. 1–12; 'Scientific realism and naturalistic epistemology', in P. Asquith and R. Giere (eds), *PSA 1980*, vol. II, pp. 613–662; 'The current status of scientific realism', in J. Leplin (ed.), *Scientific Realism*, (Berkeley: University of California Press, 1984), pp. 41–82; 'Lex orandi est lex credendi', in P. Churchland and C. Hooker (eds), *Images of Science* (University of Chicago Press, 1985), pp. 3–34.

[7] Boyd, 'Lex orandi est lex credendi', p. 4.

[8] Boyd, 'Realism, underdetermination and a causal theory of evidence', p. 9.

[9] The procedure in (i) and (ii) might be the same: realism *motivates* a method by showing that it is likely to produce success, while on at least one view of explanation to show that success is to be expected is to *explain* it.

Theory, Evidence and Non-experimental Criteria: Phenomena can evidentially support a theory only if *explained* by it, but explanatory power is appraised by comparing the *structure* of a theory with previously accepted theories. For instance, a theory only *explains* a phenomenon if it accounts for that phenomenon by postulating a process that is relevantly similar to processes that are postulated by previously accepted theories (eschewing universal forces, for example[10]). So considerations of explanatory power, based on intertheoretic plausibility considerations, count as evidential. On the realist view, it is easy to see why this should be so: only theories that are *plausible in the light of background theoretical knowledge* should be constructed and considered as candidates for confirmation by experimental evidence. This radically reduces the infinite pool that are consistent with any finite body of evidence. Anti-realists see explanatory power as a (non-evidential) *pragmatic* consideration, but how could pragmatic considerations contribute to the instrumental reliability of our standards of theory choice? If they do do so, empiricist claims that theories are underdetermined by evidence must be *false*.[11]

Experimental Design: Typically, theories are tested under experimental conditions in which they are most likely to fail, if they are false. These conditions are identified by comparing the *causal structure* of the theory under test with other theories that postulate (theoretically) relevant processes: we would not consider a theory to have been *tested* were there *another* theory that provides an alternative explanation of a successful outcome for the theory under test.[12]

Cumulativity and Retention: Commitments which hang over from our previous acceptance of now-refuted theories constrain theory construction: we consider only that 'small handful' of theories which (partially) preserve the ontology and mechanisms of previous theories. This would not make sense unless we regarded the success of the old theories as indicators of their successful reference and their approximate truth.[13]

Reference and Univocality: Realists take it that the theoretical terms of successful theories refer to theory-independent entities.

[10] See Boyd 'Realism, underdetermination and a causal theory of evidence', pp. 7–9.

[11] The evidential underdetermination thesis is formulated by Boyd as the claim that evidence for a theory is evidence of *equal force* for its empirical equivalents (ibid., p. 2).

[12] See Boyd (ibid. pp. 10–11) for a detailed example.

[13] See Putnam, *Meaning and the Moral Sciences*, p. 21, and Boyd, 'scientific realism and naturalistic epistemology', p. 619, for detailed examples.

Thus where two theories invoke (say) atoms, it makes perfect sense to apply the claims made by one theory about atoms to the atoms invoked by the other theory. This would mean, for instance, that it is *unacceptable* for the two theories to make incompatible claims about atoms, even where they cover different domains. Again, the realist view—that success indicates successful reference—motivates the strategy and explains its success.[14] On an anti-realist view, in contrast, theoretical entities are but players in empirically adequate yet *fictional* stories. Assuming univocality on *this* view would be as much of a mistake as worrying that a character plays the violin in one novel, while in another—unrelated—work of fiction, someone with the same name is tone deaf.

In *The Current Status of Scientific Realism*, Boyd challenges the chief anti-realist traditions—empiricism and (social) constructivism—to *explain* the instrumental reliability of the above methods. Take empiricism first: non-experimental criteria of theory acceptance like unity, simplicity and explanatory power are *non-evidential* according to the empiricist, and could not contribute to a theory's future fulfilment of *evidential* criteria of theory acceptance such as logical consistency and empirical adequacy. The fact that 'pragmatic' considerations *do* so contribute is inexplicable to the empiricist. Now the theory-dependence of appraisal criteria has long been recognised in the *constructivist tradition*. It is no surprise to the constructivist if our best theories fit 'reality': the theory-dependent methods are construed as procedures for the social *construction* of reality. The problem for the constructivist account of method is the *manifest* reliability of the technological products of scientific advance:

> It is ... evident that theory dependent technological progress (the most striking example of the instrumental reliability of scientific *methods* as well as theories) cannot be explained by an appeal to social construction of reality. It cannot be that the explanation for the fact that airplanes, whose design rests upon enormously sophisticated theory, do not often crash is that the paradigm *defines* the concept of an airplane in terms of crash resistance.[15]

Realism, in contrast, can both motivate the methods *and* explain their instrumental reliability. According to the realist, background theories suffuse the very methods of science. If such 'collateral' theories are (approximately) true, it is easy to see how they contribute to the likelihood that new theories—constructed and appraised in line with the methods—will be approximately true,

[14] See Boyd, ibid. section 2.4.
[15] Boyd, *The Current Status of Scientific Realism*, p. 60.

and therefore instrumentally reliable. The planes we make reliably fly because they have been designed with the help of reliable theories. We have reliable theories because we appraise theories in the light of criteria of appraisal that, although theory dependent, are reliable. Our criteria of appraisal are reliable because we have background theoretical *knowledge*.

Do the premises of the explanationist methodological argument, even if true, allow us to infer the conclusions intended by their authors? Among the premises were two claims: (i) that realism is *embodied* in scientific practices; (ii) that these practices have met with instrumental success. Realism is then inferred as the best explanatory (meta-)thesis. Opponents of realism object to this argument at every possible juncture. Firstly, one can dispute whether realist strategies of theory construction like ontological conservatism[16] or selection for unity and explanation[17] generally *are* followed, and where followed, whether they *have* typically issued in success. In effect, one might doubt whether the 'reliability' of realist methods *requires* explanation. Secondly, one might provide a motivation for supposedly 'realist' methodological principles that appeals only to *empirical adequacy* as aim and epistemic attitude. These first two objections attack the premise—methodological realism—directly.[18] One might, however, admit the realist motives but attack the *explanation* (as van Fraassen has *also* done), supplying an alternative explanation of the success of the methods.[19] Lastly, one can object to the very *structure* of the argument: an inference to scientific realism as the best explanation of methodological realism.

The structural objection runs as follows: realists present theories—literally construed—as the best explanations of the phenomena they explain. Anti-realists object to inferences-to-the-best-explanation where the best explanation in question 'traffics in unobservables' (as Lipton has put it[20]), for the following reasons.

[16] See L. Laudan, 'A confutation of convergent realism', in J. Leplin (ed.), *Scientific Realism* (Berkeley: University of California Press, 1984), pp. 218–249, at pp. 235–239.

[17] See Cartwright, *How the Laws of Physics Lie*, pp. 44–53.

[18] For instance A. Fine, 'The natural ontological attitude', in J. Leplin, *Scientific Realism*, pp. 83–107, at pp. 87–89: on the 'small handful' strategy; and van Fraassen, *The Scientific Image*, pp. 83–87, on the conjunction argument.

[19] See van Fraassen, *The Scientific Image*, pp. 93–94 and section 3, below.

[20] See P. Lipton, *Inference to the Best Explanation* (London: Routledge, 1991).

Explanations that traffic in unobservables are underdetermined by empirical evidence. In making an inference to the best explanation, we must choose the best explanation from among a necessarily restricted but historically contingent pool of theories: those that *happen* to have been constructed. The point is that there will *always* be further explanatory theories that we have not considered. If providing a good explanation were to provide a reason to *believe*, we would need (i) a reason to think that the *correct* explanation must be among those that have been *proposed*,[21] and (ii) a reason to think that what *we* count as the best explanation is the likeliest to be true. We have neither. If the anti-realist objection to realism rests on the evidential underdetermination thesis, it is easy to see why anti-realists *fail* to be impressed with the classic defence of realism in which the truth (or successful reference) of a theory is presented as the best explanation of its predictive success. The best explanation again 'traffics in unobservables'—this time the (approximate) truth of a theory's claims about unobservable entities and processes—and is as subject to underdetermination as its ground-level counterpart.[22] The explanationist methodological argument has the same objectionable inference at its heart, arguing as it does to realism as the best explanation of the reliability of realist methods, and so *it* need not trouble the anti-realist.[23] Compare the realist explanatory claims at different levels:

Level 0: The best explanation of the behaviour of cloud chambers is provided by the existence of electrons as described in our best theories.

Level 1: The best explanation of the success of our predictions of the behaviour of cloud chambers is provided by the (i) successful reference to real entitles and (ii) approximate truth of our best theories about electrons.

Level M: *Aspirational version*: The best explanation of our successful use, in learning how to manipulate the world to our advantage, of methods that are appropriate to discovering truths about real entities is provided by the (i) successful reference and (ii) approximate truth of our best theories.

Semantic-epistemic version: The best explanation of the

[21] 'We can watch no contest of the theories we have so painfully struggled to formulate, with those no one has proposed' (van Fraassen, *Laws and Symmetry* (Oxford: Clarendon Press, 1989), p. 143).

[22] Even supposing that the truth of a theory *can* explain anything.

[23] See Fine, 'The natural ontological attitude'; Laudan ('A confutation of convergent realism', pp. 242–243) puts the same argument in famously trenchant style.

reliability of methods that presume our best theories to be referentially successful and approximately true is provided by the (i) successful reference and (ii) approximate truth of our best theories.

Anti-realists deny that a claim about explanation like that at level 0—even if accepted—could launch an inference. The arguments at levels 1 and M are of the same form, and will not convince there, either: they are *parasitic* on the argument level 0. Evidently the main issue is *not* the methodological claim, but the argument's central inference. No matter *how* good our evidence for methodological realism, we cannot launch an argument for realism that would convince anti-realists from *this* pad.[24] If this sounds like acquittal on a technicality, I think there is a substantive point here: to the anti-realist—convinced of the underdetermination thesis—it is axiomatic that the world would look *just the same* if all our presently-accepted theories and background assumptions were *false*, and some other—empirically adequate, yet perhaps unborn—set *true*. This *includes* the history of advances in science, and the application of theory-dependent methods in the achievement of those advances. Thus the anti-realist could just *re-apply* the underdetermination thesis to the realist's explanation at the methodological level.[25]

2. The Vindicationist Argument

In 'The Bohr atom, models and realism', R. I. G. Hughes presents a methodological argument for scientific realism that replaces the inference to the best explanation with a direct *vindication*. Central to Hughes' argument is the *use* of physical models in probing the content of theories: he makes a distinction between 'surface' mod-

[24] Such arguments could appeal only to those who *already* entertain realist intuitions. Now Lipton (*Inference to the Best Explanation*, chapter 9) sees some value in arguments that have this property: convinced realists might justifiably use such arguments to preach to those who accept the premises and rules of inference. This is a sad end for an argument of high ambition.

[25] If the realist objects that the evidential underdetermination thesis has been *refuted* by Boyd, the anti-realist could point out that Boyd's argument showed only how *pragmatic* considerations dissolve the *practical* underdetermination 'problem'. Boyd's argument that these considerations are truly *evidential* employed something like the inference to the best explanation at level M. That, of course, begs the question at issue here.

els, which merely 'map... the phenomenal terrain' and those that (in Hughes' analogy) are more like subway maps for those without a ticket, positing unseen *underground* connections between isolated surface phenomena. A historical example of this distinction is then given: Kepler's 'purely kinematic' model of planetary motions, contrasted with Newton's dynamical model.[26] Now surface models are comparatively rare and less successful, which indicates to Hughes that the enhanced explanatory and predictive power enjoyed by the Newtonian (compared to the Keplerian) model arises from its embedding within a wider dynamical theory that is interpreted realistically.

In keeping with his paper's title, Hughes' central example is the development by Bohr of the atomic model that bore his name. Bohr constructed his atomic model within a generally classical framework, his chief explanatory problem being a theoretical one: how could a 'sun and planet' picture of the atom like Rutherford's be stable with respect to mechanical disturbance? Bohr *was* able to explain this, but at the expense of assuming *ad hoc* that only a *countable* infinity of orbits around the nucleus were available to an electron: in classical mechanics there is no reason why a *continuum* of allowed trajectories should not be possible. From the start, Bohr used his model to construct qualitative explanations and predictions covering a wide range of physical and chemical phenomena. Unfortunately, the *quantitative* versions of these theories turned out to be unforthcoming (the calculations were too difficult or problematic) or else empirically inadequate. However, at an advanced stage in the writing of his 1913 papers, Bohr added a new feature to the theory: a mechanism for the emission of radiation, and therefore an account of atomic spectra (as late as January 1913, Bohr explicitly *excluded* such a development). When Bohr performed the relevant calculations for the hydrogen atom, he was able to predict the gross structure of the absorption spectrum for atomic hydrogen—including several previously unobserved series of spectral lines—and the Rydberg constant for hydrogen R_H to within 7% of its empirical value.

[26] R. I. G. Hughes, 'The Bohr atom, models and realism', *Philosophical Topics*, **18** No. 2 (1990), pp. 71–84, at p. 74. This contrast is open to a historical objection: Kepler *did* embed his model in a wider cosmological framework (see T. S. Kuhn, *The Copernican Revolution* (Cambridge, Mass.: Harvard University Press, 1957), pp. 209–219). Nor was his model kinematic in the technical sense: his second law was derived by considering the action on the planets of a *driving force* originating in the sun. Perhaps these aspects have been ignored because only the *surface* of his model has lived on as an *advance*.

Hughes makes much of the fact that it was predictions that were *unforeseen* at the time of the model's construction that were—famously—corroborated. Now Bohr would not, perhaps, have so eagerly extended the explanatory domain of his model from the intended problem area (atomic stability) to include unrelated phenomena (spectroscopy), had there not been the possibility that it captured some aspect of reality. However, it would have been entirely reasonable to expect such a model—*realistically interpreted*—to furnish explanations of apparently unrelated phenomena that were, at the time, thought to be atomic in origin.

Now Hughes rejects the traditional argument from the empirical success of a model to the existence of the entities appearing therein: such an argument must appeal to an inference of the form 'same phenomenal structure, therefore same internal structure'. This would be unjustifiable given that two entities with different internal structures may exhibit the same behaviour: 'To take an obvious example, behind the same software can lurk many different kinds of hardware.'[27] A conclusion that *can* be drawn, he claims, is the existence of the *subject* of the model—the atom—rather than its purported constituents (the electrons and nuclei). The argument runs as follows: the practice of model-building presupposes the existence of its subject. Model-building is sometimes a successful scientific endeavour. Where successful, model-building is vindicated. The building of Bohr's model was successful, and was therefore vindicated. So, therefore, is the assumption on which this activity is predicated: to wit, (in this case) the existence of atoms. Hughes eschews the *explanationist* argument, preferring what I have called the *vindicationist* variant:

> I am proposing a simple criterion of justification, applicable in all spheres of practical reason. If by adopting a certain practice we are led to success, then in this case the practice is justified. If not, not.

Now we see why the existence of the *constituents* is not inferred: the activity of model-building does not presuppose *that*:

> The practice we are looking for is that of building a constitutive model. It involves two things: (1) assuming that an entity exists; (2) modelling its behaviour in a particular way. If the model is successful, then both elements of the practice have been justified. The justification extends no further than the actions described.

[27] Hughes, 'The Bohr atom', note 26.

He concludes that

> what has been justified is precisely the assumption that a partic-
> ular kind of entity, exhibiting a certain kind of behaviour,
> exists.[28]

Now the *strength* of the realist thesis that would be justified by this
argument is unclear. There are two possible readings of the phrase
'exhibiting a certain kind of behaviour': the weak reading makes
the conclusion uncontroversial (but not an especially *realist* one),
but on the strong reading it is obviously false. Hughes baulked at
the inference to a realist thesis about the inner structure of a model
from its external—observable—structure. Perhaps what has been
justified, then, is the existence of something that exhibits the
observable behaviour of the model. The weak version is this: let us
infer that something exists and call it 'atoms', where 'atoms'
invokes only the regularities that realists cite as evidence for the
existence of something *behind* the phenomena.[29] Thus Hughes
quotes Suppe:

> A theory has as its intended scope a natural kind of class of phe-
> nomenal systems... In propounding a theory, one commits one-
> self to the existence of the phenomenal systems within the
> theory's scope.[30]

Would this commitment be peculiar to the realist? Surely not: the
conclusion would be *banal*, because knowledge that 'atoms' existed
would not imply our possession of any knowledge that could be
subject to empiricist objections. For to say that the *empirical sub-
model* of the model of a theory has a counterpart in reality is just to
say that the theory is *empirically adequate*. Hughes stresses the
realist credentials of his conclusion, indicating a stronger reading:

> Simply put, ... the assumptions at work in Bohr's theory, and jus-
> tified by its success, are that atoms exist, that they are stable, that
> those of a given element are uniform in size, etc—and that only
> an anti-realist axe-grinder would describe matters otherwise.[31]

Now there is no evidence that Bohr *himself* saw any distinction
between the *non*-essential (internal) structure of his model, and the
existential assumptions that were presupposed by its very con-

[28] Hughes, 'The Bohr atom', p. 81.

[29] In terms of the *semantic* view of theories, the weak view of what is
justified is that part of reality is correctly described by the empirical *sub-
model* of the successful model.

[30] Suppe, quoted in Hughes, 'The Bohr atom', note 28.

[31] Hughes, 'The Bohr atom', p. 81.

struction. Nor do we have any reason to think that the facts about atoms—stability, uniformity of size—that have been established to Hughes' satisfaction were presupposed by Bohr's model constructing. Maybe what Bohr presupposed was that atoms *under his theoretical description* existed, but it has been clear for some time that no set of entities *do* have all the required properties.[32] These, however, are relatively minor matters, for there is a more serious problem: whatever its strength, the conclusion cannot follow from the premises of the argument, for its trades on a conflation.

The conflation concerns two notions of justification. Hughes' 'simple criterion of justification' acts in the field of *practical reason*, and justifies a *practice*—in this case the building of models and using them *realistically*—as efficient means to some such desirable end as instrumental success. This criterion is eminently reasonable: we are surely justified in pursuing our ends via means that have previously satisfied those aims. In the argument, however, the justification is transferred to the realistic presuppositions of the practice, and is read as *epistemic* justification. Even supposing the transfer to be possible—that having a belief or making a presupposition is the kind of thing that can be vindicated with respect to an aim as if it were a voluntary *action*—it is the wrong *kind* of justification. In practical reasoning, we work with an *instrumental* notion of rationality: we judge the instrumental rationality of a practice by the success it (reliably) allows, that is, by its connection with certain *consequences*. So the *vindication* has to be with respect to some desired *aim*.[33] That is why Hughes' argument cannot work: when we achieve our aims by actions that only make sense because we have certain beliefs, it does not follow that the presuppositions are true.[34] We could *at best* conclude that we had

[32] Unlike the causal arguments (see above, n.2), we are given no special feature that picks out some subset of our theoretical claims about atoms for special attention.

[33] Giere in 'Scientific rationality as instrumental rationality', in *Studies in the History & Philosophy of Science*, **20** (1989), pp. 377–384, seems to offer an argument with a similar structure: he raises the question of realism, giving a methodological answer (in DNA research, a realist programme yielded the greatest 'payoff'). What is unclear is whether we are meant to conclude that therefore DNA exists and has the structure attributed to it by Crick and Watson, or just that, with hindsight, their methods yielded the greatest payoff.

[34] Unless, of course, we know more about the *mechanism* by which the means are appropriate to the ends. If we knew the mechanism, and it required that the presuppositions were *true*, we could say that the truth of the presuppositions was the *best explanation* of our achieving our aims. That, however, is another argument.

to have those beliefs in order to achieve those aims by those means, not that we must have those beliefs *simpliciter*.[35]

3. Methodological Realism

So far, my conclusions have been predominantly *negative*: neither the explanationist nor the vindicationist attempts to fill out the methodological argument for realism straightforwardly support their intended conclusion. In this final section, however, I will defend the common premise of these arguments (methodological realism, henceforth: MR), remaining *neutral* on scientific realism (henceforth: SR). MR is the claim that the adoption—by scientists—of realist aims, methods and inferences is (or has been) central to their construction and acceptance of (what turned out to be) successful theories. It is a small step from this to a *recommendation* that scientists adopt realist methods. This raises two groups of issues. Firstly, what *is* the logical relation between MR and SR? Does MR make *sense* as a recommendation outside the realist view of science?[36] Secondly, there is MR itself: I noted at the end of the first section that some non-realist critics (Fine and Laudan, for instance) chose to attack the *premises* of the methodological arguments (MR, that is). So *is* MR in fact a plausible claim when applied (for instance) to the history of science?

Consider first whether we can defend MR independently of a broadly realistic view of the aims, methods and products of science. The problem is this: even though we have seen the failure of some of the arguments for SR from MR, isn't MR in some sense *incoherent* given an anti-realist view of science? According to SR the aim of science is truth, so that success in science is the possession of theories that are (approximately) true. Not only that, but SR claims those aims to be *achievable*: in all probability we advance toward their fulfilment with every new predictively suc-

---

[35] As, ever, Duhem provides an apposite quote: 'Chimerical hopes may have incited admirable discoveries without those discoveries embodying the chimeras that gave birth to them. Bold explanations which have contributed greatly to the progress of geography are due to adventurers who were looking for the golden land—that is not a sufficient reason for inscribing "El Dorado" on our maps of the globe' (P. Duhem, *The Aim and Structure of Physical Theory*, translation by P. Wiener of his *La Theorie Physique: Son Objet, Sa Structure* of 1914 (Princeton University Press, 1954), pp. 31–32).

[36] J. Leplin, 'Methodological realism and scientific rationality', *Philosophy of Science*, **53** (1986), pp. 31–51, defends a similar position to MR, but does not consider its logical relation to scientific realism.

cessful and explanatorily powerful theory. MR is then almost trivial: it recommends the adoption of methods that are *appropriate* to realist aims. In the conjunction of SR and MR, then, we have an agreeable confluence of aims and methods. Anti-realists, in contrast, *deny* that we have good reasons to think any but the *empirical* claims our best theories make are true. Our aims should be limited to what we can achieve. Now MR commends methods that are appropriate to *realist* aims, methods that also presume that those realist aims can be (and in fact have been) partially achieved by previously successful theories. Surely it is folly to adopt methods that are reasonable only on the assumption that we can know what—in principle—we *cannot* know.

This, however, is too quick: all that the harmony between MR and SR can show is their consistency (or perhaps that MR is a natural consequence of SR), but the methodological arguments seek to establish SR on the basis of MR. To the earlier criticisms of these arguments I would like to add the claim that SR is in fact *independent* of MR. This I will argue for by showing that MR can be motivated in either realist *or* anti-realist views of science.[37] If the consistency of MR and SR has already been established, it remains only for me to remove the air of paradox from the adoption of MR within an *anti*-realist framework. This proves to be surprisingly easy, for in van Fraassen's empiricist account of science we have a ready-made candidate.

First consider again the three realist theses set out in the introduction, concerning: (i) aims; (ii) semantics; and (iii) epistemic attitude. How might these theses, adopted by scientists, affect their practice? Realists often claim that, at the level of practice, realist aims counsel a search for theories that *explain*, rather than merely save the phenomena. The semantic and epistemological claims work together: a realistic construal of our best theories directs our interest to the consequences of conjunctions of theories covering disparate domains, and also licenses univocality assumptions and intertheoretic plausibility considerations (see section 1). The epistemological claim rationalizes the retention of (portions of) previously successful—but now refuted—theories.

Compare this with van Fraassen's constructive empiricism: 'Science aims to give us theories that are empirically adequate; and acceptance of a theory involves as belief only that it is empirically adequate'. However, van Fraassen *accepts* the realist's semantic claim: 'After deciding that the language of science must be literally understood, we can still say that there is no need to believe good

[37] In effect I will take a cue from elementary logic: if I exhibit models of (A & B) and (A & ⌐B), I have shown that B is *independent* of A.

theories to be true, nor to believe *ipso facto* that the entities they postulate are real'.[38] So van Fraassen rejects realist theses (i) and (iii), but accepts (ii). Now for van Fraassen, epistemology is unlike totalitarian codes of law: it allows what it doesn't specifically prohibit, rather than prohibiting what it doesn't specifically allow.[39] The content of the realist's *belief* in a theory is logically stronger than the constructive empiricist's acceptance of it,[40] but belief in the extra content is not *irrational*, for the contents of the two beliefs are *empirically equivalent*, the extra realist belief could not make us more vulnerable to empirical surprises, and that is what counts.[41]

This covers only *static* (epistemic) features of acceptance, but the happy effects of realist commitment outlined by Boyd and Putnam concerned *dynamic* features of science: how would our previous acceptance (in van Fraassen's sense) of a theory affect the shape of theories that we build and accept in *future*? Here van Fraassen turns to the pragmatics of theory acceptance: acceptance of a theory may involve *not only* the belief that it is empirically adequate, but also a commitment to a research programme, and to framing future explanations in its terms. Acceptance of a theory may also involve immersing ourselves in its world picture, letting it constrain the vocabulary *and* grammar of our theoretical discourse. Thus: 'to some extent, adherents of a theory must talk just as if they believed it to be true.'[42] A survey of van Fraassen's answers to particular realist arguments concerning scientific practice underlines this *methodological indistinguishability* of acceptance and belief.[43] However, one does get the curious feeling that the actions of the realist just get *re-described* in empiricist terms rather than explained or motivated—here we are talking, theorizing and predicting as if we are realists, but with our anti-realist hearts pure.

In any case, a more interesting consequence of van Fraassen's separation of evidential and pragmatic aspects of theory choice

[38] Van Fraassen, *The Scientific Image*, pp. 11–12.

[39] See van Fraassen, *Laws and Symmetry*, pp. 171–176.

[40] In terms of the semantic view of theories: belief in the empirical adequacy of a theory is just the belief that part of one of its models—the empirical sub-model—corresponds to part of the world.

[41] See B. van Fraassen 'Empiricism in the philosophy of science', in P. Churchland and C. Hooker (eds) *Images of Science* (University of Chicago Press, 1985), pp. 245–309, at p. 255.

[42] *The Scientific Image*, p. 202.

[43] See his discussion of Putnam on conjunctions and Boyd on experimental design in *The Scientific Image*, Ch. 4.

(and therefore of epistemology and methodology) allows him to endorse the realist search for *explanatory* theories. The problem is this: the history of science is littered with highly successful theories for which their creators sternly held out, seeking explanation, when there were extant theories that *saved* the phenomena. Answering Feyerabend's charge that a search for empirical adequacy *alone* might therefore hinder scientific progress[44]—the search for explanation might have 'paid off handsomely'—van Fraassen argues:

> Paid off handsomely, how? Paid off in new theories we have more reason to believe empirically adequate. But in that case even the anti-realist, when asked questions about *methodology* will *ex cathedra* counsel the search for explanation! We might even suggest a loyalty oath for scientists, if realism is so efficacious.[45]

This throws a different light on the discussion of acceptance. Faced with some feature of scientific practice whose plausibility appears to depend on acceptance of realist theses (i) to (iii), the anti-realist can *either* construct an alternative *empiricist* rationale *or* accept the apparent realistic commitment of the method, but endorse the method as conducive to empiricist aims. But does not the second strategy collapse into the first? No: the force of Feyerabend's argument was that the search for explanation might not be so fruitful were it not pursued *as an end in itself.* There is a parallel in ethical theory: utilitarians endorse the acceptance *by others* of a non-utilitarian rule if its compliance utility is higher than that of a corresponding utilitarian rule.[46] In both ethical and methodological cases, the possibility of such an endorsement is a direct consequence of the consequentialist's separation of motivation and appraisal. Now van Fraassen accommodates only the realist *motives* by this mechanism, but anti-realists who reject *other*

[44] Feyerabend, in 'realism and instrumentalism: comments on the logic of factual support' in M. Bunge (ed.), *The Critical Approach to Science and Philosophy* (New York: Free Press, 1964), pp. 280–308, cites the impetus to developments in dynamics provided by difficulties encountered by the Copernican system—*realistically construed*—against the background of the prevailing Aristotelian dynamics.

[45] *The Scientific Image*, p. 93.

[46] 'Thus a Utilitarian may reasonably desire, on Utilitarian principles, that some of his conclusions should be rejected by mankind generally; or even that the vulgar should keep aloof from his system as a whole, in so far as the inevitable indefiniteness and complexity of its calculations render it likely to lead to bad results in their hands' (H. Sidgwick, *Methods of Ethics* (London: Macmillan, 1877), pp. 448–449).

elements of the realist position—concerning semantics and patterns of commitment—could endorse *their* adoption in the same way.[47]

In conclusion, both the realist and anti-realist might accept MR's recommendation: in both settings, the plausibility of MR depends crucially on claims about past and present scientific practice, although the appraisals will differ with respect to aims and the efficacy will be explained by different mechanisms. In Boyd's account, the plausibility of the realist view of science as a whole depends on its explanatory power with respect to the reliability of realist methods. For an anti-realist the plausibility of MR might instead depend on a meta-induction: have realist methods *in fact* contributed to our possession of empirically adequate theories? So now we turn to the second set of issues.

Fine has criticized Boyd's realist motivation for the small handful strategy. His chief claims are that: (i) realism *fails* to explain (and therefore motivate) the 'small handful' strategy, whereas (ii) instrumentalism *succeeds* in providing such a motivation. At any one time, so the (realist) argument goes, instead of the infinite host invoked by underdetermination arguments, only a *small handful* of possible theories are considered by scientists: those that are related to the most successful refuted theory in certain special ways. This is reasonable because the old theory was approximately true, and our attention might fruitfully be confined to those theories which stand in some correspondence relation or have a 'family resemblance' to it. If one is a realist, one can expect that the new theory will inherit the happy characteristic of approximate truth. The explanationist claim is that SR both rationalises the strategy and provides the only explanation of its success. Fine counters that SR doesn't, in fact fulfil either of these claims. The explananda are three: the small size of the handful, the family resemblance, and the success of the strategy. On the first question, the realist would still be faced with an infinity of choices. On the second question, SR again fails, because we can't infer the approximate truth of the *old* theory from its previous empirical success: *that* is as suspect as any other realist inference. Therefore the *new* theory's success cannot be explained on the basis of its transferred approximate truth: there might be none to transfer. Worse, even if the old theory was

[47] There need be no hint of self-deception here: the appraiser stands outside the community—scientists—whose beliefs are appraised on their effects. Another analogy is to the functionalist explanation of religion in primitive societies (see Y. Elkana, 'A programmatic attempt at an anthropology of knowledge', in E. Mendelshon and Y. Elkana (eds), *Sciences and Cultures* (Dordrecht: Reidel, 1981), pp. 42–44).

approximately true, the element of truth that it captured might not be passed on to the new theory, such is the logic of approximate truth. The third explanandum requires no explanation, because it is *false*: the small handful strategy too often *fails*.

On the first two questions, Fine claims, the methodological instrumentalist—one who claims that theories are constructed with purely pragmatic virtues in mind—has an explanation that enjoys the virtues of the realist's, while dispensing with the dubious allusion to approximate truth. Consider the 'small handful' part: constructing theories that conform to known empirical constraints is difficult, but we might as well focus on the few theories that satisfy the following pragmatic considerations. Firstly, it is quite reasonable to keep the highly confirmed bits of an old theory, while tacking on some new bits to cover any evidence that refuted it. Retention of familiar and tractable mathematical structures has the same rationale, which explains the family resemblance. Thus the conservatism of the small handful strategy is easily explained given that empirical adequacy is the aim. If it often fails, this reflects the 'trial and error' nature of the exercise. SR, meanwhile, has 'struck out'.[48]

Now Fine does not distinguish between SR and MR, and some of his criticisms of SR need not threaten MR: it doesn't matter whether *we* conclude that the old theory was approximately true, for what counts is whether the *scientist* thinks that it is (in some specified respect). Meanwhile, the instrumentalist version of the 'small handful' strategy is too permissive because it falls foul of Feyerabend's objection: so much the worse for subsequent science if the *Copernicans* had been happy with one of Fine's botched-together theories. Nor is there any general reason to think that the mathematical tractability could be a very important factor: Heisenberg's matrix mechanics was famously *in*tractable, but was hailed by Pauli—the arch-critic of the (*ad hoc*, but empirically adequate) old quantum theory—as the long-awaited advance.

Laudan has a more confrontational criticism: he just *denies* that new and old theories are generally related in ways that are amenable to a realist rationale.[49] Cartwright, taking as her examples the quantum mechanical models that are actually *used*, urges that models—the vehicles of scientific achievement—rarely bear the right relations to the fundamental theories in which they are purportedly embedded. The theories don't explain the models, and the success of the models won't support the theories as *factual*

[48] Fine, 'The natural ontological attitude', p. 89.
[49] See Laudan, 'A confutation of convergent realism'.

claims:[50] the realist tells the wrong story about the majority of cases of model-building. two responses are possible here: either (i) get into an argument about numbers, or (ii) limit the realist claims to the 'commanding heights' of theorising. The rationale is as follows: when applying quantum mechanics to lasers, numerical accuracy is important: model-building will be a messy and pragmatic business. However, think of the contrast with the initial introduction of the *same* theory in 1925: would we really have counselled Pauli to be content with the old quantum theory? Instead of *numbers*, the methodological realist would choose to argue historical counterfactuals: would we have had heliocentrism, Bohr's atom or matrix mechanics without stern demands for explanation?

[50] See Cartwright, *How the Laws of Physics Lie*, pp. 100–127.

Quantum Technology: Where to Look for the Quantum Measurement Problem

NANCY CARTWRIGHT*

1. Introduction

This paper, I am afraid, advocates the philosophy of technology without actually doing it. It can best be seen as a plea for the philosophical importance of technology; in this case, importance to one of the most widely discussed problems in philosophy of physics—the measurement problem in quantum mechanics. What I want to do here is to lay out a point of view that takes the measurement problem out of the abstract mathematical structure of theory, where we discuss questions about unitary operators or conditions for the disappearance of certain inner products supposed to represent interference terms, and locate it elsewhere. Where is the measurement problem? Answer: It had better be found in the quantum technology or it is not to be found at all. My view in many respects follows ideas I have learned from Willis Lamb.[1]

The evolution of states is supposed to be governed by the Schroedinger equation. The measurement problem arises when we become convinced that there are a range of cases in which we no longer want to assign the state dictated by the Schroedinger equation but want to assign a different state instead. Why do we want to assign this second state? We call the problem "the measurement problem" and this canonical example is a good illustration. You

*My thanks to Jordi Cat for help with this paper.

[1] Willis Lamb (1969), 'An operational interpretation of nonrelativistic quantum mechanics', *Physics Today* **22**, pp. 23–28; W. Lamb (1986), 'Quantum theory of measurement', *Annals, New York Academy of Sciences*, **480**, pp. 407–416; W. Lamb, 'Theory of quantum mechanical measurements', in *Proceedings of the 2nd International Symposium on the Foundations of Quantum Mechanics: In the Light of New Technology*, ed. M. Namiki *et al.*, (Japan Physical Society, pp. 185–192); W. Lamb, 'Classical measurements on a quantum mechanical system', 1987, *Nuc. Phys. B (Proc. Suppl.)* **6** (1989), pp. 197–201; W. Lamb, 'Suppose Newton had invented wave mechanics' (unpublished, 1993); W. Lamb, 'Quantum theory of measurement: three lectures', lectures delivered at the London School of Economics in July 1993.

can't measure a quantum system without coupling to it. If the coupling is modelled quantum mechanically then at the end of the interaction the apparatus and system are in a composite quantum state that is a superposition across eigenstates of the apparatus pointer observable. But the pointer, we know, points in a definite direction. It has, to macroscopic accuracies, a definite position.

So, how do we get from here to a *problem*? Basically by assuming that all true descriptions are renderable as quantum descriptions. The pointer has a position. We used to have a classical physics that treated positions—systems with position were assigned classical states and the behaviour of these states was encoded in classical mechanics. If we are going henceforth to use only quantum mechanics, all these descriptions must go, and we will have to find some analogue in quantum mechanics for the pointer position that is so well treated by classical mechanics. The best candidate seems to be an eigenstate of an operator we dub 'the pointer observable'. But this quantum state is incompatible with the Schroedinger-evolved state. Hence the measurement problem.

My proposed strategy, consistent with the kind of theoretical pluralism I advocate in general, is not to succumb to the quantum takeover. The world is rich in properties—they're all equal citizens here. We long ago learned that there are properties like positions and momenta which are well represented by classical mechanics. The discovery that there are also features—primarily of the micro-world—that are well represented by quantum states and well treated by quantum theory does not in itself give us reason to throw out those properties that have been long established. So my claim is this: There are both quantum and classical states, and the same system can have both without contradiction. It is important here that I say *classical* states, not quantum analogues of classical states. There is no contradiction built in because there is no *theory* of the relation between quantum and classical characteristics. As with all cases of genuine theoretical pluralism, what we have to do is look for what connections there are and where they are. That is what the technologist is doing daily. And that is why I say that we must hunt the measurement problem in quantum technology.

2. The Relation between Quantum and Classical Descriptions

I begin with the familiar observation that the Schroedinger equation guarantees a deterministic evolution for the quantum state function. Nevertheless quantum mechanics is a probabilistic theory. What are the probabilities probabilities of? One answer is that they are probabilities that various 'observables' on the system possess certain allowed values. This is the answer associated with the *ensemble interpretation* of quantum mechanics. It is widely rejected on the grounds that results like those of the 2-slit experiment show that observable quantities do not have 'possessed values' in quantum systems. The more favoured alternative is that the probabilities are probabilities for certain allowed values to be found when an observable is measured on a quantum system.

I reject both of these answers. I propose instead that the probabilities (when they exist at all) are probabilities for classical quantities to possess particular values. Quantum mechanics is primarily a theory about quantum states, how they evolve and how they interact. Occasionally a quantum analysis will allow us to predict facts about the classical state of some system. These predictions are (for the most part) probabilistic, and there is no way to eliminate the probabilistic element. But there is no guarantee that a quantum analysis will yield such predictions. There is no universal principle by which we infer classical claims from quantum descriptions. This is rather a piece-meal matter differing from case to case.

The fundamental idea I want to urge, following my interpretation of Lamb's ideas, is that some systems, perhaps all, have quantum states, some have classical states, and some have both. The presupposition is that macro-objects have classical states and micro-objects have quantum states. Perhaps macro-objects have quantum states as well. Macro-interference effects could give us one reason to assign a quantum state to macro-objects in certain kinds of situations. Sketches of micro-macro interactions like Schroedinger's cat or the Von Neumann theory of measurement have also been thought to provide macro quantum states. A third commonly accepted reason for assigning quantum states to macro-objects relies on standard procedures for modelling quantum interactions. It is supposed that (i) given any two systems with quantum states, the composite of the two must itself have a quantum state represented in the tensor product of the spaces representing the two separately; and also that (ii) all interactions between systems with quantum states are quantum interactions, representable by an interaction Hamiltonian on the tensor product space. Hence step by step the quantum state of a macro-object is built up out of the quantum states of the systems that make it up.

I myself do not find any of these reasons compelling, and especially not the second and third, which make claims about what kinds of treatments quantum mechanics can in principle provide without actually producing the treatments. But that is no matter. For the point is that there is no automatic incompatibility between quantum and classical states even when they do describe the same system. The relation between quantum and classical states follows no simple pattern; perhaps contradictions will be unearthed in one case or another, but they are not automatic.

Characteristics of physical systems are classically represented by analytic functions onto real numbers. In classical statistical mechanics, for instance, these quantities are functions of the basic dynamical variables, position and momentum, and hence they are defined on the two-dimensional phase space that the position and momentum values define. On this phase space physical states are represented by probability densities. Position and momentum are also basic dynamical variables in quantum mechanics, although in this case they are represented mathematically by noncommuting Hermitian operators on a Hilbert space. It is on this complex vector space that quantum physical states are defined and can be associated with a statistical operator, the density matrix, that represents probability distributions.

Despite some formal analogies, however, quantum and classical quantities fall under different mathematical characterizations and there is no general rule that enables us to represent any classical quantity with a quantum operator and vice versa. We have to make a variety of attempts to formulate such rules but they have all proven problematic. In my opinion this is entirely the wrong strategy. Though quantities represented in classical and quantum physics are mutually constraining (in different ways and different forms), they are different quantities exhibiting behaviour that is formalized differently in the two theories. We should not expect to be able to represent classical quantities in the quantum formalism, nor the reverse. In particular we should not expect to be successful in the familiar attempt to represent macroscopic classical quantities by a set of commuting operators in quantum theory.

There is one widely accepted assumption that offers a simple pattern of connection between quantum and classical descriptions. That is the GENERALIZED BORN INTERPRETATION. I think that the claim of this interpretation is unwarranted, though very specific instances of it may not be. (Born's own identification of $|\Psi|^2$ with detection probabilities in scattering experiments is a good example.) The generalized Born interpretation supposes that linear Hermitean operators

$$O = \Sigma e_i \, |\phi_i> <\phi_i|$$

on a Hilbert space represent "observable" quantities on the systems represented by vectors in that Hilbert space. The interpretation says that

$$\text{Prob}_\Psi (O=e_i)=<\Psi|\phi_i><\phi_i|\Psi>$$

As before there are two views about the meaning of '$O = e_i$': (i) 'O possesses the value e_i', or (ii) 'The value of e_i will be found in a measurement of O'. Both readings have well-known problems. I think these problems arise because we are trying to find a simplistic route that does not exist for generating usable predictions and for testing the quantum theory. As is often remarked, the 'measurement' that appears in this interpretation is no real measurement, but something more like what a genie riding on the back of the particle would read out without needing any instruments.[2] With quantum mechanics we can, and do, make predictions about what happens in real measurement situations and in real devices such as spectroscopes and lasers and SQUIDS. But the analysis of all such situations is difficult and the grounds for the inferences from the quantum analysis to predictions about the behaviour of the devices are highly various and context dependent.

Lamb's own semi-classical theory of the laser provides a good example. The theory assumes a classical electromagnetic field in interaction with the laser medium. The field induces a 'dipole moment' ($<p>=<er>$) in the atoms of the medium; the dipole expectation is identified with the macroscopic polarization of the medium; this in turn acts as a source in Maxwell's equations. Sargent, Scully and Lamb[3] provide the diagram reproduced in Figure 1.

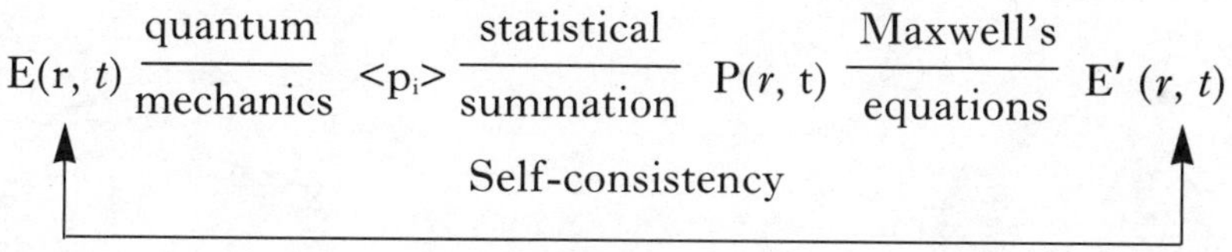

$$E(r, t) \xrightarrow[\text{mechanics}]{\text{quantum}} <p_i> \xrightarrow[\text{summation}]{\text{statistical}} P(r, t) \xrightarrow[\text{equations}]{\text{Maxwell's}} E'(r, t)$$

Self-consistency

Figure 1. Electric field E assumed in cavity induces microscopic dipole moments (p_i) in the active medium according to the laws of quantum mechanics. These moments are then summed to yield the microscopic polarization of the medium $P(r, t)$, which acts as a source in Maxwell's equations. The condition of self-consistency then requires that the assumed field E equal the reaction field E' (Sargent *et al.*, *Laser Physics* p. 96).

[2] See James Park's measurement₁, and measurement₂ in J. Park (1969), 'Quantum theoretical concepts of measurement', *Phil. of Science* **35**, pp. 205–231.

[3] M. Sargent III, M. O. Scully and W. Lamb, Jr., *Laser Physics* (Reading, MA: Addison-Wesley, 1974).

In the semi-classical theory the first step, linking the field and the dipole expectation, appears as a causal interaction between classical and quantum characteristics. In the step marked 'statistical summation' we see an identification of a quantum quantity, the suggestively named 'dipole expectation', with a purely classical polarization. I put the term 'dipole expectation' in scare quotes to highlight the fact that from the point of view I am defending we should not think of it as a genuine expectation. I have rejected the assumption that every linear Hermitean operator represents some quantity (or, inaptly, some 'observable') which has or will yield values whose probability distribution is fixed by the generalized Born interpretation. Without the Born interpretation there are no automatic probabilities, and no expectations either. The identification of <er> with the macroscopic polarization is a case in point. Here the 'interpretation' proceeds by identifying a quantum inner product with the value of a well-known classical quantity. The identification is guided by a powerful heuristic: the quantum electric dipole oscillator. We are told by Sargent, Scully and Lamb, as we are in many texts on laser physics, that quantum electrons in atoms 'behave like charges subject to the forced, damped oscillation of an applied electromagnetic field' (p. 31). This charge distribution oscillates like a charge in a spring, as in Figure 2.

The discussions of the oscillator waver between the suggestive representation of <er> as a genuinely oscillating charge density and the more 'pure' interpretation in terms of time-evolving probability

(a) P=0   (b) P≠0

Figure 2. Stimulated emission and dipole oscillators. (a) Spherical electron charge distribution of hydrogen atom, indicated by plus sign with centre of distribution at nucleus (1s energy state). (b) Application of electric field shifts distribution (centre at—.) relative to positively charged nucleus. Subsequent removal of field produces oscillating distribution. that is, an oscillating dipole like a charge on a spring (Sargent *et al.*, *Laser Physics*, p. 32).

amplitudes. From the standard point of view this should be troubling; without a generalized Born interpretation looming in the background I see no reason to be upset. The oscillator analogy provides a heuristic for identifying a quantum and a classical quantity in the laser model. The identification is supported by the success of the model in treating a variety of multimode laser phenomena—the time and tuning dependency of the intensity, frequency pulling, spiking, mode locking, saturation, and so forth.[4]

Consider now how problems are usually generated for views that assign a new state to the measuring device after measurement. There are two ways. First using Von Neumann's theory of measurement it is supposed that the measurement interaction takes the composite into the state $\Sigma c_i \, \phi_i \, \Psi_i$ for some set of 'pointer position' eigenstates $\{\Psi_i\}$. When the interaction ceases each pointer must have a possessed value for position and thus be in one of the eigenstates Ψ_i. Since under the generalized Born interpretation pointer position 'i' will occur with probability $|c_i|^2$, by observing the distribution of the pointer positions we can infer the values of the $|c_i|^2$. The problem arises when we consider the implications of the assumption that each pointer state is Ψ_i for some i. In that case, it is argued, the state of the mated system in each case must be ϕ_i since under the generalized Born interpretation the probability is zero for the pointer to have value i and the system to have some different value e_j (j ≠ i) for the measured observable O. So for each individual composite the joint state must be $\phi_i \, \Psi_i$; the ensemble of these composites will be then in the mixed state $\Sigma |c_i|^2 |\phi_i><\phi_i| \, |\Psi_i><\Psi_i|$. This contradicts the original assumption that the ensemble was in the pure, or non mixed, state $\Sigma c_i \phi_i \Psi_i$.

When we do not try to render pointer states as quantum states (like Ψ_i) but rather as classical states, no such contradiction arises—at least not yet. I say 'not yet' because I take it that in each new analysis the question remains open. There are ways of inferring quantum states from information about classical states, as we are now discussing. Nothing I have said, or will say, guarantees that no situations will arise in which the classical information we can get in that situation and the inferences we take to be legitimate will produce a new quantum state for the apparatus and system that is inconsistent with the supposition that the Schroedinger-evolved superposition continues to be correct. But finding these situations, undertaking the analysis, conducting the experiments

---

[4] For a second example see my more extended discussion in 'Where in the world is the measurement problem', *Philosophia Naturalis*, (forthcoming).

and drawing this surprising conclusion is a job that remains to be done. This is the job of *hunting the measurement problem*.

The second way to generate a measurement problem is to look at the time-evolved behaviour of system and apparatus to see if the predictions generated from the superposition contradict those from the second state assignment. Again a contradiction is possible: We may find some interactions where the quantum-evolved state implies classical claims and these claims contradict the information contained in the classically-evolved states. But such analyses are very uncommon and very difficult. We virtually never give a serious quantum analysis of the continuing interactions of the measuring apparatus. To product a contradiction we not only must do this but we must do this in a situation where we can draw classical inferences from the final quantum state. Again the job in front of us is not to solve the measurement problem *but to hunt it*.

3. Philosophical Remarks[5]

I close with two brief philosophical discussions. First, what happens if we reject the generalized Born interpretation? How then do we interpret quantum mechanics? I begin by pointing out that usually the discussion of the measurement problem presupposes a strongly realist view about the quantum state function. People like me who are prepared to use different incompatible state assignments in models treating different aspects of one and the same system are hardly troubled by the contradictions that are supposed to arise in special contexts of measurement. But it is puzzling why quantum realists should be calling for interpretation. For those who take the quantum state function seriously as providing a true and physically significant description, the quantum state should need no interpretation. There is no reason to suppose that those features of reality that are responsible for determining the behaviour of its microstructure must be tied neatly to our 'antecedent' concepts or to what we can tell by looking. Of course a fundamental property or state must be tied causally to anything it explains. But laying out those ties need look nothing like an interpretation.

Interpretations as standardly given in quantum mechanics attempt to exhaust what the quantum state is by reference to classical properties. The result is almost a reductive definition, except that the characterization is not in terms of a set of possessed classical properties but rather in terms of dispositions to manifest these

[5] This section is taken directly from my 'Where in the world is the measurement problem', *Philosophia Naturalis* (forthcoming).

properties. Behaviourism is an obvious analogy, and I think an instructive one. The distinction between behaviour and mental states is drawn in the first instance on epistemological grounds. Whether or not these epistemological grounds are reasonable in the case of behaviourism, the analogous distinction in quantum mechanics is a mistake. We know very well that we can measure quantum states, although as I have argued, there is no universal template for how to go about it as the generalized Born interpretation presumes. So quantum mechanics is epistemologically in no worse position than classical mechanics. Clearly for the behaviourist more than epistemology was involved: The distinction took on an important ontological status. Behaviour was thought to be the stuff really there; mental-state language merely a device for talking in a simple way about complicated patterns of behaviour. This is clearly not the kind of position that quantum realists want to find themselves suggesting.

My second remark is a very brief one about a very big philosophical topic. Why do standard accounts use quantum states rather than the more obvious choice of classical states that I urge, following Willis Lamb? I suspect the answer can be traced to what I would take to be a mistaken view about what realism commits us to. Let us grant that quantum mechanics is a correct theory and that its state functions provide true descriptions. That does not imply that classical state ascriptions must be false. Both kinds of descriptions can be true at once, and of the same system.

We have not learned in the course of our work that quantum mechanics is true and classical mechanics false. At most we have learned that some kinds of systems in some kinds of situations (e.g. electrons circulating in an atom) have states representable by quantum state functions, that these quantum states evolve and interact with each other in an orderly way (as depicted in the Schroedinger equation), and that they have an assortment of causal relations with other non quantum states in the same or other systems. One might maintain even the stronger position that all systems in all situations have quantum states. I think this is a claim far beyond what even an ardent quantum devotee should feel comfortable in believing. But that is of no matter here. Whether it is true or not, it has no bearing one way or another on whether some or even all systems have classical states as well, states whose behaviour is entirely adequately described by classical physics.

One reason internal to the theory that may lead us to think that quantum mechanics has replaced classical physics is reflected in the generalized Born interpretation: We tend to think that quantum mechanics itself is in some ways about classical quantities, in which case it seems reasonable for the realist to maintain that it rather than classical theory must provide the correct way of treating these. But that is not to take quantum realism seriously. In developing quantum theory we have discovered new features of reality that realists should take to be accurately represented by the quantum state function. Quantum mechanics is firstly about how these features interact and evolve and secondly about what effects they may have on other properties not represented by quantum state functions. In my opinion there is still room in this very realistic way of looking at the quantum theory for seeing a connection between quantum mechanics operators and classical quantities and that is through the correspondence rule that we use as heuristic in the construction of quantum mechanical Hamiltonians. But that is another long chapter. All that I want to stress here is my earlier remark that quantum realists should take the quantum state seriously as a genuine feature of reality and not treat it as an instrumentalist would, as a convenient way of summarizing information about other kinds of properties. Nor should they insist that other descriptions cannot be assigned besides quantum descriptions. For that is to suppose not only that the theory is true but that it provides a complete description of everything of interest in reality. And that is not mere realism, but imperialism.

But is there no problem in assigning two different kinds of descriptions to the same system and counting both true? Obviously there is not from anything like a Kantian perspective on the relations between reality and our descriptions of it. Nor is it a problem in principle even from the perspective of the naive realist who supposes that the different descriptions are in one-to-one correspondence with distinct properties. Problems are not there just because we assign more than one distinct property to the same system. If problems arise, they are generated by the assumptions we make about the relations among those properties. Do these relations dictate behaviours that are somehow contradictory? The easiest way to ensure that no contradictions arise is to become a quantum imperialist and assume there are no properties of interest besides those studied by quantum mechanics. In that case classical descriptions, if they are to be true at all, must be reducible to those of quantum mechanics. But this kind of wholesale imperialism and reductionism is far beyond anything the evidence warrants. We must face up to the hard job of practical science and continue to figure out what predictions quantum mechanics can make about classical behaviour.

In the account of quantum mechanics that I have been sketching the relations between quantum and classical properties is complicated and the answer is not clear. I do not know of any appropriately detailed analysis in which contradiction cannot be avoided. Lamb's view is more decided. So far we have failed to find a single contradiction, despite purported paradoxes like two-slit diffraction, Einstein Poldolsky-Rosen or Schroedinger's cat: 'There is endless talk about these paradoxes, Talk is cheap, but you never get more than you pay for. If such problems are treated properly by quantum mechanics, there are no paradoxes.'[6]

Still, many will find their suspicions aroused by these well-known attempts to formulate a paradox. If so, our task is to find the measurement problem, not as many of us have supposed, to solve it. I advise them suspicious, then, to turn from their mathematical abstractions and instead to look hard at quantum technology. For only there they might find a measurement problem that does not rest on the shaky metaphysics of monopoly, reduction and take-over.

[6] Willis Lamb 'Quantum theory of measurement: three lectures', lectures delivered at the London School of Economics in July 1993, p. 3.

Welcome to Wales: Searle on the Computational Theory of Mind

ROGER FELLOWS*

In a recent book devoted to giving an overview of cognitive science, Justin Lieber writes:

> ...dazzingly complex computational processes achieve our visual and linguistic understanding, but apart from a few levels of representation these are as little open to our conscious view as the multitudinous rhythm of blood flow through the countless vessels of our brain.[1]

It is the aim of hundreds of workers in the allied fields of Cognitive Science and Artificial Intelligence to unmask these computation processes and install them in digital computers.

Professor Kevin Warwick of the University of Reading says that within fifteen years there will be machines appreciably more intelligent than any human being and, echoing John McCarthy, he foresees: 'a machine-based intelligent environment, and we're just what we would regard as animals within that'.[2]

The image of the computer or robot endowed with genuine mentality resonates deeply in the collective psyche of late twentieth-century Western culture, and that image often has the dark overtones hinted at by Professor Warwick and made explicit in films such as *Westworld*. And no wonder. The computer is the most complex technology ever devised by man, and we hold it up as a mirror to our own souls.

I suppose that most researchers in the field of what John Searle has termed 'Strong AI' would distance themselves from Professor Warwick's apocalyptic vision, but many would think that the Socratic injunction 'know thyself' is best obeyed by focusing on computational theories of the mind, and not on, for instance, Skinnerian behaviourism or Freudian psychology.

In this paper I am going to discuss two thought experiments. These are the usual well known ones: Turing's Imitation game and Searle's Chinese Room argument. I shall suggest that, properly

* I am grateful for comments made by Margaret Boden, Nancy Cartwright, David Cooper and Graham Macdonald.

[1] *An Invitation to Cognitive Science* (Oxford: Blackwell, 1991), p. 114.
[2] Reported in *University Life*, Vol. 1, No. 1 (Jan. 1994), p. 228.

understood, Turing's test does provide a sufficient condition for the ascription of mentality to machines. Naturally the caveat 'properly understood' bears all the load of the argument. I almost feel the need to apologize for raising, yet again, the spectre of Searle's Chinese Room argument, which has repeatedly exasperated workers in the field. I do not think that what I have to say here is entirely original, but I hope there will be some interest in the way in which I formulate the reason why this argument should be rejected. Turing, it will be recalled, argued that fluency in some natural language, English for example, is the appropriate test for deciding whether a computer can think: *the fluency to be judged by the interrogator*; and Searle's argument is designed to show that a computer could be a Turing test passer without understanding English. Here is another way of saying it: Searle believes that he can show that the print-outs of a digital computer are not assertions, but marks upon paper which the interrogator construes as assertions.

I also wish to discuss Searle's new argument to the effect that computational theories of mind are infected with the homunculus fallacy. This argument does not rest upon a thought experiment, and, in my opinion, is the more interesting of the two.

Misunderstandings of what Searle takes himself to have demonstrated with the parable of the Chinese Room are widespread. In a recent introductory work on Artificial Intelligence,[3] the author interprets Searle's view as that we are prepared to attribute cognitive states only to things which are made from the same sort of stuff as ourselves: 'Searle's only basis for denying that robots think is that they are not made of flesh and blood.'[4]

This view has been referred to as Carbon chauvinism. But it is not Searle's own view. Searle makes it quite clear,[5] that it might be possible to build a thinking machine out of non-biological components: silicon chips for example. Searle says:

'I have not tried to show that only biologically based systems like our brains can think. Right now those are the only systems we know for a fact can think, but we might find other systems in the universe that can produce conscious thoughts, and we might even come to be able to create thinking systems artificially. I regard this issue as up for grabs.'[6]

[3] A. Garnham, *Artificial Intelligence An Introduction* (London: Routledge & Kegan, Paul, 1988).

[4] Ibid., p. 231.

[5] Particularly in his article 'Is the brain's mind a computer program?', *Scientific American* (Jan 1990).

[6] Ibid. p. 21.

Searle's claim is that digital computers cannot be endowed with mentality in virtue of running a computer program. The central processing unit of a digital computer is in essence a Universal Turing Machine. So Searle is claiming that there is no program or collection of programs alone which could transform a physical device equivalent in power to a Universal Turing Machine into a genuinely thinking system.

Human beings are, in the jargon, intentional systems, and we may even be physical systems equivalent in power to a Universal Turing Machine, but we do not possess minds in virtue of running computer programs: minds do not stand to brains as the computer program stands to the hardware of the digital computer. (As we shall see later, Searle's new argument aims for the conclusion that the statement that the brain is, in essence, a Universal Turing Machine is meaningless.)

Many workers in the field have responded to Searle by pointing out that a thinking machine is one which is appropriately programmed and which bears the right causal relations to the world, but Searle believes that this objection fails to blunt the force of the Chinese Room argument.

The Chinese Room argument is a thought experiment which can be described very simply. You are asked to imagine that you are locked in a room with a slot in one wall. In the room you have a number of baskets containing slips of paper on which are printed Chinese symbols. You also have a book of rules which are of the form: 'If a string of shapes ****** comes through the slot on a slip of paper then assemble a string of shapes @@@@@@ on a slip of paper and push it back through the slot.' You identify the Chinese symbols by shape because you lack any knowledge whatsoever of the Chinese language. The book of conditional instructions is written in English which you do understand. Suppose that outside the room there are people who understand Chinese and who are asking the 'system' = (room + symbols + rule book + manipulator) questions in Chinese and interpreting the slips of paper pushed through the slot as answers to the questions in the Chinese language. The rule book is the computer program, you are the computer and the baskets full of symbols are the data base. The people who wrote the book are of course, the programmers. In his Reith lectures, Searle says:

'Suppose the programmers are so good at designing the programs and that you are so good at manipulating the symbols that very soon your answers are indistinguishable from a native Chinese speaker.'[7]

Searle's point is that you, in the room, with the basket of symbols and rule book, collectively pass the Turing test for understanding Chinese: but you do not understand Chinese, and nothing else in the room does either. So a system cannot think in virtue of implementing a computer program.

Now I, in common with many others, do not find this argument compelling. I agree that it seems to me that I can imagine the Chinese room passing the Turing test in the manner described by Searle: but the inference from 'I can imagine that p is is possible' to 'p is possible' is invalid. The invalidity of this inference has been dubbed 'The Cartesian fallacy' by some philosophers because Descartes is supposed to have shown that the thing which things is not spatially extended by appealing to the premise: 'I can imagine or conceive that I exist and that no bodies exist'. Now, I do not want to go into the question of whether or not Descartes did commit this fallacy; and, as it happens, I do not think that he did, but there is no doubt that others have, as the sad catalogue of circle-squarers and angle-trisectors makes clear. A more dramatic illustration is provided by the demise of Hilbert's programme at the hands of Godel. These are examples of people imagining that p, and its turning out that p is *logically* impossible.

There are also instances of people imagining things which turned out to be *physically* impossible. A statement's physical possibility is defined in terms of its being consistent with the laws of nature and, unless we know all the laws of nature, we may not be able to tell, at any given time, whether a statement is physically possible or not. Before the Special theory of Relativity, for instance, physicists believed that, if you advanced towards a ray of light with velocity v, which was coming towards you head on with velocity c, then, by the theorem of the addition of velocities, the combined velocity would be $c + v$. I want to discuss whether the Chinese Room thought experiment is an instance of logical or physical impossibility. Many of the philosophers and psychologists whose work I have read on this topic believe that Searle simply ignores real time constraints in setting up the thought experiment; but we can, I think, say a little more.

Syntax is the study of the conditions under which sentences are either well- or ill-formed: that is, syntax studies the rules which govern the stringing-together of words and phrases into sentences.

[7] John Searle, *Minds, Brains and Science* (London: Penguin, 1984), p. 32.

Searle on the Computational Theory of Mind

When one person utters a sentence which is syntactically well-formed in the hearing of another, the latter knows that it is well-formed within constraints such as attentiveness and memory (for instance, the hearer might forget the first part of the sentence if it is very long). Similarly, fluent speakers of a language can tell when they are confronted, *even for the first time*, by syntactically ill-formed strings such as 'Value the virtue when'. This suggests that we should suppose that a grammar of a natural language is *recursive*: which means that the set of rules should be sufficiently powerful to tell both when a sentence is grammatical and when it is not. This after all is a capacity possessed by humans. Because it is a human capacity, then if we can get the syntax right, we may be justified in supposing that we are providing an explanation, in some sense, of what is going on when people utter and comprehend sentences as grammatical and ungrammatical, unless we countenance the view that nature works by magic. Notice also that the set of sentences e.g. of English, and the set of ill-formed sentences, are both countable: otherwise, we should not need a recursive grammar of English, and we could simply give two finite lists.

The above is, very clearly, a highly compressed account of work in the tradition of Chomsky.[8] I am relying on this tradition for my objections to the Chinese Room argument. *We shall see below that Searle's new argument, if it were successful, would undercut my objection to Searle's thought experiment.*

My argument can be stated as follows: since the list of questions which could be fed through the slot in the wall of the room forms a countable set, it follows that the set of answers is also countable. But this must mean that the rule book must contain countably many conditional instructions. No book with a finite number of pages can contain a countable number of instructions. Therefore the Chinese Room experiment is impossible. To guard against a possible misunderstanding, I am not arguing that any one Chinese interrogator of the room can utter an infinite number of sentences,

[8] As people acquainted with Chomsky's work will know, the early work in Transformational Grammar has been replaced by a principles-and-parameters approach. The language faculty is now conjectured to contain two parts. The first part consists of a nearly language-invariant computational procedure and the second part consists of a (highly) variant lexicon. A very important difference between the older approach and contemporary linguistic investigations is that the traditional grammatical constructions—noun phrase, verb phrase, prepositional phrase, etc., are viewed 'as taxonomic artifacts, their properties resulting from the interaction of far more general principles'. See N. Chomsky, 'Language and nature', *Mind*, Vol. 104 (1995), p. 413.

nor that, if the Chinese population came to the room one at a time to ask every question that they could think of, the union of such questions would form a countable set. The point is that the rule book itself must list all the questions which *might* be asked: and it cannot do so if it contains only a finite list of conditional instructions. An objection which may be made here is that there are many questions which could be put to me that I cannot answer either, so the rule book need contain only a finite set of rules: but this objection rests upon a misunderstanding of the argument. While it is true that I do not know, like everybody else, the answer to all the questions which might be put to me, I can answer that I do not know the answer.

Is Searle's thought experiment logically or physically impossible? I believe that it is physically impossible. There *is* a logical impossibility concealed within the Chinese Room. The logical impossibility is that a book with a finite number of pages cannot contain the countable set of conditionals necessary to enable the symbol—shuffler to relate questions to answers within a room with a finite spatial volume. Such a book would need to have a countable number of pages. But this impossibility is the consequence of the claim that natural languages are to be characterized recursively.

If it is a necessary truth that a human language must have a recursive characterization then it follows that Searle's thought experiment is a logical impossibility. Now, it might be objected that I have misdescribed Searle's thought experiment. I have supposed that the book in the Chinese Room contains a countable number of instructions of the form *if x is inputted then output y* (where 'x' and 'y' range over sentences of Chinese), and that this simple switching architecture does not do justice to the complexity of the situation in the room. Searle does speak of the person in the room having baskets of symbols, and selecting symbols from different baskets to assemble an output string as the 'answer' to a given input string. So while it is true that there are a potentially infinite number of questions that can be fed into the room, the person in the room can assemble the answers to any of these questions from a finite set of baskets containing a finite collection of symbols and a rule book containing a finite list of conditional instructions. In a word, the instruction book in the Chinese room contains rules of recursion.

Suppose we grant this. I would then claim that the Chinese Room thought experiment is physically impossible, and is no more a challenge to the computationalist than is the claim that I can imagine myself travelling faster than light a challenge to relativistic physics. There is a particularly effective way of demonstrating this claim. One may purchase from the University of Stanford, California, a computer-based introduction to computability theory for the Macintosh.[9]

If one runs even simple programs by utilizing the computational speed of the computer, it is overwhelmingly obvious that no human being could imitate the computation in real time. The whole point of the Chinese Room thought experiment was to show that the person in the Chinese Room could push out strings of symbols in real time, that the Chinese interrogators of the room would interpret as answers to questions. But if the person in the Chinese Room has to *consciously* carry out the recursive computations (much as I might plough my way through a very long truth-table for a sentence of propositional logic) then the people outside the room are usually going to be waiting a very long time for their answers. And the person in the room *does* have to perform his or her computations consciously: since to say that he or she could perform them unconsciously would be to go along with some version of the computational theory.

Searle says that his thought experiment is designed to remind us of the conceptual truth that syntax is not sufficient for semantics. Since programmes are syntactical, his result follows independently of musings on people in rooms shuffling Chinese symbols. But the computationalist never supposed that the programme *simpliciter* has a mind: any computer or robot which is to be treated as an agent, an entity which is the bearer of P and M predicates, to put it in Strawsonion terms, will require transducers which convert physical effects into discrete input. Other transducers will be required to convert computational output into action.

Now Searle has considered this reply and, in the Reith lectures, he had this to say:

> If the robot moved around and interacted causally with the world wouldn't that be enough to guarantee that it understood Chinese? Once again the inexorability of the semantics—syntax distinction overcomes this manoeuvre. ... You can see this if you imagine I am the computer. Inside the room in the robot's head I shuffle symbols without knowing that some of them come in to me from television cameras attached to the robot's head and others go out to move the robot's arms and legs. As long as all I have is a formal computer program, I have no way of attaching any meaning to any of the symbols. And the outside world won't

[9] J. Barwise and J. Etchemendy, *Turing's World 3.0 An Introduction to Computability Theory* (Stanford, California, 1993).

help me to attach any meanings to the symbols unless I have some way of finding out about the fact.[10]

There is something wrong about Searle's response here. When I look at a table the input into my brain is *not* a string of symbols but a stream of physical events which get transduced into neuro-physiological events in the brain. If the computationalists are right, the brain has the causal power to convert the incoming phys-ical signals for input into computer programs and then to compute these programs rapidly. So there is a crucial disanalogy here.

Before moving on to discuss Searle's more recent argument I would like to add my support for those who think that conversa-tional fluency in a natural language is a sufficient test for the pres-ence of mentality. Let me remind you that Turing only hides the machine from the interrogator in his thought experiment because he wishes to eliminate any suspicion of silicon chauvinism; there is no other reason why the computer or robot should be hidden from view. And Donald Davidson, in a recent and acute discussion of the imitation game,[11] has urged that the experiment is flawed pre-cisely because of the imposition of a 'veil of ignorance' between questioner and answerer. The robot must be brought into view, because, says Davidson, we need to be satisfied that the device or artificial agent uses singular terms and predicates correctly and that it knows the truth conditions of its sentences. We need, in short, to observe the device interacting with both ourselves and the environment. Davidson would require that we are satisfied that the device not only says true things about the world, but that it says true things about the things which we say to it.

Davidson also argues that, for it to be the case that we are satis-fied about the foregoing, it will be necessary that the device resem-bles ourselves to a certain extent and that it has an appropriate his-tory. For example, we will want to be able to grasp a machine ana-logue of gesticulation, and so on. We will also want to grasp to some extent the route by which the device learned to ascribe par-ticular predicates to objects; I say 'to some extent', because there is still so much we do not understand about human language-acqui-sition. I do not think that there is anything in these reflections which most computationalists would disagree with, and no one supposes that the Turing test passer in this more relaxed sense is on the near horizon. Our knowledge of natural language is still

[10] *Minds, Brains and Science*, p. 34.
[11] 'Turing's Test', in Newton Smith, Viale and Wilkes (eds.), *Modelling the Mind* (Oxford University Press, 1990).

woefully inadequate: linguists have not yet succeeded in recursively characterizing English at the syntactical level; to say nothing of the semantic and pragmatic dimensions of language. It is the complexity of the tasks which face research workers in AI which makes Keith Gunderson's parody of Turing's original test so utterly beside the point.[12] In essence, Gunderson argues that, if I were to put my foot through a hole in a wall, and I failed to distinguish correctly between the event consisting of a person stepping on my toe, and the event consisting of a rock falling on my toe, I would have to conclude that rocks can step on toes!

I turn now to John Searle's recent argument against computational models of the mind. The Chinese Room argument was designed to answer the question 'Is the mind a computer programme?' in the negative, and the recent argument is aimed at returning a negative response to the question 'Is the brain a digital computer?' Some computational theorists will not discern two questions here but one; however, what is of interest is that Searle has an argumentative strategy which is independent of the Chinese Room. Searle calls the view that the brain is a digital computer 'Cognitivism': mental states, which are identical to neurophysiological ones, are computational states. This, says Searle, defines a research programme. We try to find out what programmes are being implemented in the brain, by programming computers to run the same programs. If the computer then passes the Turing test, we have evidence that the brain is running the same programmes, subject, as ever, to the under-determination of theory by data.

I shall start by repeating a story told by Richard Taylor[13] in the context of his discussion of the Teleological Argument: it is a nice way into Searle's argument. Imagine you are on a train and, looking out of the window, you see a pattern of white stones on the hillside which form the shape: WELCOME TO WALES. Do you thereby have reason to believe that you have arrived in Wales? You do, but only if you further believe that the stones have been intentionally placed there in order to give people the information that they have arrived in Wales. Suppose you knew that the stones had rolled down the hill after a large rock had fractured due to extremes of weather: ordinary physical forces had conspired to produce the accidentally arrived-at shape WELCOME TO WALES. In this case, it would be wrong to think that you had evidence you had arrived in Wales because it would be wrong to treat

[12] Keith Gunderson, in 'The imitation game', A. R. Anderson (ed.), *Mind and Machines*, (Englewood Cliffe, NJ: Prentice Hall, 1964).

[13] In his book *Metaphysics* (Englewood Cliffs, NJ: Prentice Hall, 1963).

the stones as constituting a piece of English syntax. But, whether
the shape WELCOME TO WALES is, or is not, an English sentence, is not intrinsic to the pattern itself: it depends upon the
extrinsic intentions of the people who places the stones there, if
any did, and upon your ability to recognize the pattern of stones as
an English sentence. You will have wrongly read syntax into the
picture if the pattern of stones is the product of blind physical
forces. Syntax, says Searle, is not intrinsic to physics: it is always
observer-relative.

Now, on the usual textbook definition, computation is, as Searle
remarks, defined in syntactical terms. So the observer-relativity of
syntax and the standard definition of computation entail that you
cannot discover computation in physics, although you can assign a
computational interpretation to a physical process and, in particular, you could choose to regard the brain as a digital computer.
Searle says:

> The point is not that the claim 'The brain is a digital computer'
> is false. Rather it does not get up to the level of falsehood. It
> does not have a clear sense. You will have misunderstood my
> account if you think that I am arguing that it is simply false that
> the brain is a digital computer. The question 'Is the brain a dig-
> ital computer?' is as ill-defined as the questions 'Is it an abacus?'
> 'Is it a book?', 'Is it a set of symbols?', 'Is it a set of mathemati-
> cal formulae?'[14]

Searle argues in consequence that all computational models of cog-
nition tacitly commit the homunculus fallacy. Consider a digital
computer. Very roughly, a high level language such as COBOL is
compiled or translated into the machine language which is the only
language the computer "understands", and which consists of
sequences of 0's and 1's, represented by voltage levels in the hardware. It is at this level that the homunculus is discharged. But
Searle argues:

> To find out is an object really is a digital computer we do not
> actually have to look for 0's and 1's etc; rather we just have to
> look for something that we could *treat as* or *count as* or *could use
> to function* as 0's and 1's... So we might think that we should go
> and look for voltage levels. But Block tells us that 1 is only 'con-
> ventionally' assigned a certain voltage level. The situation grows
> more puzzling when he informs us further that we do not need

[14] John Searle, 'Is the brain a digital computer?', Presidential Address,
delivered to the American Philosophy Association, California, March 30
(1990), p. 35.

to use electricity at all, but could use an elaborate system of cats and mice and cheese to make out gates in such a way that the cat will strain at the leash and pull open a gate which we treat as if it were an 0 or 1.[15]

Again,

> For real computers, ..., there is no homunculus problem, each user is the homunculus in question. But if we are to suppose that the brain is a digital computer, we are still faced with the question 'And who is the user?'[16]

If I have got Searle right, he purports to confront the cognitivist with a disjunctive dilemma: either the brain is a digital computer because there are people around who propose to interpret it thus, in which case cognitivism is not explanatory; or, if people did not exist, then the claim that the brain is a digital computer is senseless, in which case it is likewise unexplanatory. Both disjuncts of the dilemma rest upon the fact that syntax is extrinsic to physics. In consequence, Cognitivism is powerless to provide genuine explanations for what goes on in the brain. As Searle puts it: 'In the brain, intrinsically, there are neurobiological processes and sometimes they cause consciousness. But that is the end of the story.'[17]

Searle says that, for real computers, it makes sense to ascribe computations to them since we are around to do the ascribing. Also, he maintains that part of what is involved in calling an object a computer has to do with the conventional assignment of 0's and 1's to voltage levels in the devices. A cheap debating point here would be to remark that, in the absence of intentional beings like ourselves, nothing would *actually* be a computer. Suppose the entire human race disappeared tomorrow but various computers continued to number-crunch or whatever. Would it follow that the disappearance of the human race entailed the disappearance of computers? Searle would say, I think, that in the absence of people there would still be devices going through various physical processes, which we, if we were to reappear, would interpret as computers. This response seems a plausible one because silicon-based computers are not natural kinds; so if there were no people there would be no computers *but only in the sense that there would be no objects of a certain kind answering to human interests*, just as there would, in this sense, be no tables or washing machines.

[15] Ibid. p. 25.
[16] Ibid. p. 29.
[17] Ibid. p. 36.

Roger Fellows

It seems to me that Searle's argument does not do justice to the fact that there are many explanatory levels which are not intrinsic to physics. Consider again the pattern of stones forming the 'sentences' *Welcome to Wales*. If the stones rolled down the hill and formed the pattern according to the erosion of the rocks and so on, then the explanation would lie in physics. However, one can imagine (albeit fancifully) a biological explanation in terms of a group of badgers having moved the stones around for the purposes of their own. The explanation for the resulting pattern is extrinsic to physics and lies within the biological domain. In general, unless one espouses the position of the hardline reductionist who insists that all genuine explanations must be couched in the language of physics, we have to acknowledge that there are many explanatory levels: the physical, the chemical, the biological, the psychological, and so on, no one of which can be reduced to the others in any interesting sense. Why does Searle suppose that it is senseless to say that the brain carries out computations? He argues that, if you take away the assigner or interpreter, then you take away the syntax. But what you actually do is to take away the theorist! By way of parallel, consider the fact that chemical theory *conventionally* assigns the symbols, 'H' and 'O' to a particular substance, so that for instance, water is identified as having a certain structure. Nevertheless it would remain true that water would equal H_2O even if Humanity disappeared off the map. I think therefore that the issue of the conventional assignment of symbols to physical processes is a red-herring; and that Searle's main thrust against the Computational theory of Mind rests upon his concept of explanation.

Most of this paper has been oppressively negative in tone; but I would like to end on a speculative note. I think that I regard it as an open question as to whether a digital computer could think, but I have a kind of Cartesian worry about what kind of thinking things they would be. If one looks through standard textbooks on AI, one finds chapters on pattern-recognition, knowledge-representation, language, thinking and reasoning: in short, there is an emphasis on the attempt to model our cognitive faculties. Suppose we accept, as I do, a representative realism which maintains that there is an ontological difference between primary ('objective') and secondary ('subjective') properties. Then it does not seem to be stretching the point too far to say that AI is concerned to model those features of the mind which traffic in the domain of primary properties: the domain of chess, syntax, geometry and so on. After-images, tastes, smells, pains etc, are not high on the AI agenda. (A student said to me last year: 'but why would anyone bother to build a machine in pain—there is enough around already'!)

Can we make sense of a device with a mind whose contents are characterized entirely by primary predicates, for instance 'x thinks that the pyramid is on top of the square'? Descartes thought that we could. He thought that human minds are characterized by primary predicates; and also by secondary ones, as in, 'x thinks the cube is yellow'. In such cases a the latter (i.e. of secondary predicates), it follows from Descartes' theory that x has either a false belief or a wrong mental representation. Descartes thought of angels as incorporeal beings which can have causal intercourse with the material world, which is how he thought of the human mind. Descartes was pressed to explain how then, if at all, a man differs from an angel. His answer was that:

> ...if an angel were in a human body, he would not have sensations as we do, but would simply be intellectually aware of the motions which are caused by external objects, and in this way would differ from a real man.[18]

For Descartes, we are distinguished from the angels by our sensory lives, and not by our abstract reasoning capacities. Perhaps AI is an attempt to build embodied angels.

---

[18] J. Cottingham, *A Descartes Dictionary* (Oxford: Blackwell, 1993), p. 13.

Acts, Omissions and Keeping Patients Alive in a Persistent Vegetative State

SOPHIE BOTROS

Introduction: The Medical and Legal Background

There are many conflicting attitudes to technological progress: some people are fearful that robots will soon take over, even per- haps making ethical decisions for us, whilst others enthusiastically embrace a future largely run for us by them. Still others insist that we cannot predict the long term outcome of present technological developments. In this paper I shall be concerned with the impact of the new technology on medicine, and with one particularly ago- nizing ethical dilemma to which it has already given rise.

A patient lies in a hospital bed. He is not in a coma. He sleeps, wakes, sleeps again. His limbs remain crooked and taut but are capable of primitive reflex movements. The touch of bedclothes can elicit a grasp reflex. He sometimes chews, scratches, swal- lows, grinds his teeth. He may even grunt and groan. The patient's family, in hope, interprets these movements as signify- ing the stirrings of renewed mental activity. But the neurological experts shake their heads. They know that the patient is not clini- cally dead because the brain stem is intact. But he has suffered the almost total disintegration of his cerebral cortex. He is thus in a state of wakefulness without any kind of awareness: he neither sees, feels nor hears. He cannot communicate in any way. He is profoundly and permanently insensate. For a person in this con- dition, the medical profession has coined an acronym: PVS.[1] The patient, usually an accident victim, is in a 'persistent vegetative state'.

Until quite recently people who had sustained this massive brain damage would not have survived the initial physical trauma which caused it. Nowadays however they can be kept alive for years; breathing unaided, they require only to be artificially fed and hydrated. The moral question therefore arises whether doctors should continue to feed and hydrate them or whether after some stipulated period ensuring that the diagnosis is certain, such as

[1] See B. Jennett and F. Plum, 'Persistent vegetative state after brain damage', *The Lancet* (1 April 1972), pp. 734–737.

perhaps a year, they may withdraw the naso-gastric tubes, with the result that the patients die from starvation.

A case recently brought to court was that of Anthony Bland, whose lungs had been crushed in a football stadium disaster, and whose doctors and parents wanted the feeding tubes removed so that he might die.[2] The case finally reached the House of Lords who ruled, largely on the grounds that it was an *omission* rather than an act of *commission*, that the withdrawal of these feeding tubes, unlike the giving of a lethal injection, was neither murder nor manslaughter.[3] That it was not a *culpable* omission depended upon two further premises: that artificial food and hydration was 'treatment', and that no treatment could medically benefit a patient in PVS.[5,6]

A number of the Law Lords however expressed concern about the *moral* underpinning of the legal distinction between act and omission upon which their judgment ultimately rested.[7] This

[2] See *Airedale N.H.S. Trust v. Bland* [1993] 2 W.L.R. 316 (Fam.D.) and (C.A.)

[3] See *Airedale N.H.S. Trust v. Bland* [1993] 2 W.L.R. 316 (H.L.) at 368E–369C per Lord Goff: 'Why is it that the doctor who gives his patient a lethal injection which kills him commits an unlawful act and indeed is guilty of murder, whereas a doctor who, by discontinuing life support, allows his patient to die, may not act unlawfully—and will not do so, if he commits no breach of duty to his patient?... the reason is that what the doctor does when he switches off a life support machine "is in substance not an act but an omission to struggle and that the omission is not a breach of duty by the doctor because he is not obliged to continue in a hopeless case"'. This publication is henceforth referred to as *H.L.*

[4] Ibid. at 372H per Lord Goff, at 378G–H per Lord Lowry and at 362C-D per Lord Keith.

[5] Ibid. at 362E per Lord Keith, at 372E–F per Lord Goff and at 386H per Lord Browne-Wilkinson.

[6] This should not be taken to imply that, for the Law Lords, culpability is not itself directly relevant to whether the withdrawal of life-sustaining treatment counts as murder (or manslaughter). On the contrary, as Lord Browne-Wilkinson makes clear at 383F, if there is a duty to treat because that is seen to be in the patient's best interests, and the patient dies as a result of the failure to treat, this omission to treat, being culpable, may count as murder (or manslaughter).

[7] Ibid. at 379D–H, Lord Lowry attributes the following tentative critique of their Lordships' judgment to a 'hypothetical non lawyer': 'The solution here seems to me to introduce what lawyers call a distinction without a difference... might it not be suggested... that this case [the withdrawal of life sustaining treatment from Bland] is, in effect if not in law, an example of euthanasia in action?' At 381D, Lord Browne-Wilkinson acknowledges that some would see the distinction between it

paper attempts to address their disquiet. It asks whether the moral doctrine of acts and omissions can plausibly bear the weight here being placed upon it.

The conclusion of our investigation will be that, though a version of this doctrine can be formulated that does, from the perspective of what is sometimes called 'ordinary' or 'common sense' morality,[8] have moral relevance, there are special features of the medical context which preclude the withdrawal of food and water from patients in persistent vegetative state counting as in this sense an 'omission'. This is, however humane, deliberate killing, and as such, prohibited at least from the perspective of ordinary morality.

This caveat however is essential if my conclusions are not to be misunderstood. I should not automatically be taken to be endorsing the standpoint of ordinary morality, or indeed any general appeal to the acts/omissions doctrine, how ever coherent such a doctrine is shown to be relative to that morality.

Indeed if anything this paper suggests reasons why we ought *not* unreflectively to endorse ordinary morality. Herein lies its practical importance. Nothing it would seem could be more practically important than establishing that our familiar system of duties and near absolute prohibitions, particularly those concerning the protection of human beings, lacks the underpinning which would per-

---

[8] I have in mind the familiar system of duties and near absolute prohibitions, particularly those concerning the protection of innocent human life, which forms the background of current legal and ethical public debate.

---

and taking active steps to end life as 'artificial'. Lord Mustill, at 388H–389A, expresses 'acute unease ... about adopting this way through the legal and ethical maze', and attributes his unease to 'the sensation that however much the terminologies may differ the ethical status of the two courses of action [withdrawing life-sustaining treatment from Bland and 'mercy killing'] is for all relevant purposes indistinguishable'. Their Lordships' difficulty seems to spring from their view that when a doctor withdraws life-sustaining treatment from a PVS patient, such as Bland, thereby allowing him to starve to death, his intention can only be to terminate the patient's life (see Lord Browne-Wilkinson at 383E). Now the charge of murder (or manslaughter) partially rests upon this very presumption of an intention to terminate life, in a case, unlike Bland's, where a doctor deliberately withholds treatment from a patient knowing that it is in his best interests, and that he is likely to die without it. Why then should a similar charge not be applicable in the Bland case? Not all writers however agree that there is such an intention in a case, such as Bland's, and I discuss the issue below, pp. 115–118.

mit practices, which many people would regard as only humane.[9] Those who look with trepidation at hospital beds being increasingly filled by such insensate patients are then at a crossroads: whether to cling to their old moral beliefs or to turn to some other moral system which, despite allowing deliberate killing, might possibly fill the gap Lord Mustill suggested had arisen between 'old law and new medicine'.[10]

I: Acts and Omissions: The General Problem

(i) A hard-nosed dismissal

For some, a swift way of dealing with the acts/omissions doctrine would be to reject it as incompatible with their more general moral approach. Thus utilitarians could argue that since only consequences ultimately matter, it is quite irrelevant whether these are produced by actions or omissions: even deliberately bringing about someone's death is not in itself worse than failing to save them. Those who hold that rights are morally fundamental, but are ascribable only to individuals capable of desire, could argue in the same way with regard to those who lack the relevant desires, since in neither case, according to these rights theorists, would any rights be violated.[11]

But obviously this hard-line approach will fail to impress those not similarly committed to the moral theories in question. Moreover it overlooks the indisputable fact that deliberately bringing about the death, even of the senile or permanently comatose, seems patently worse to ordinary moral intuition than merely refraining from saving them. More circumspect critics therefore typically acknowledge these common sense moral responses but then attempt to show that they can be accounted for

[9] See Lord Goff (*H.L.* 370C), who suggests that to allow a patient like Bland to die is 'required by common humanity'.

[10] Ibid. at 379H where Lord Lowry attributes the remark to Lord Mustill.

[11] M. Tooley develops this position in 'Abortion and infanticide', in P. Singer (ed.), *Practical Ethics* (Oxford University Press, 1988), pp. 57–86. We shall see below that most of the major discussions of acts/omissions are conducted in terms of the distinction between killing and letting die. This is unfortunate since 'killing' is ambiguous as regards whether there is an intention to bring about death, and this ambiguity obscures some key questions about the nature and justificatory role of the acts/omissions doctrine (see note 35 below). I have attempted as far as possible, and even at the cost of circumlocution, to remove this ambiguity where it arises. It is often however so deeply entrenched as to be ineradicable.

without recourse to the bare distinction between an action and an omission.

(ii) Some more circumspect criticism: 'comparison' and 'conflict' examples[12]

Two general types of example, so called 'comparison' and 'conflict' examples, are associated with this line of reasoning. In the first, two situations are envisaged which correspond in every detail except that, in one, A's death comes about directly as a result of B's action, say he pushes A's head under water, whereas in the other A's death is a result of B's refraining from saving her from drowning. We are then invited to compare the two situations in order to test the claim that B's conduct in the first situation is morally worse than in the second. The object is to isolate from the surrounding circumstances the bare fact of something's being an action rather than an omission, in order to assess whether this difference has in itself any moral import.

James Rachels offers the following now famous pair of cases:

> In the first, Smith stands to gain a large inheritance if anything should happen to his six year old cousin. One evening while the child is taking a bath, Smith sneaks into the bathroom and drowns the child. In the second Jones also stands to gain if anything should happen to his six year old cousin. Like Smith, Jones sneaks in planning to drown the child. However Jones sees the child slip and hit his head, and fall face down in the water. Jones is delighted, he stands by ready to push the child's head back under if it is necessary, but it is not necessary... the child drowns all by himself, as Jones watches and does nothing.[13]

Rachels argues that since for Jones to plead in his moral defence that he (unlike Smith) didn't drown the child, but only watched him drown, would ordinarily be regarded as a grotesque perversion of moral reasoning, it must follow that the relevant action cannot be in itself worse than the relevant omission. Rather it is extraneous factors, such as the presence of an evil motive (where by contrast sheer laziness or even panic often explains a failure to

---

[12] These helpful terms were introduced by Heidi Malm in 'Killing, letting die and simple conflicts', *Philosophy and Public Affairs* (1989); reprinted in J. M. Fischer and M. Ravizza (eds), *Ethics, Problems and Principles* (New York: Harcourt, Brace, Jovanovich, 1992), pp. 135.

[13] 'Active and passive euthanasia', *The New England Journal of Medicine*, **292** (1975), pp. 78–80.

act) or the greater certainty that death will result, that account for our typically viewing an action, such as drowning someone, as morally worse than the corresponding omission. However, once these factors are introduced into the omission version of the scenario, as in Rachels' example, it becomes clear, or so the critics will say, that the bare distinction between act and omission carries no moral weight.

It is noteworthy however (and a point that we will repeatedly return to) that precisely by making the situations similar in every respect, except whether the agent does or does not move his body in ways relevant to the act of drowning the child, Rachels has simply *assumed* that the acts/omissions distinction is this bare empirical one (as it is often called) between action and *inaction*.

More recently critics of the doctrine, whilst agreeing with Rachels that in his scenario it is just as bad to refrain from saving the child from drowning as to deliberately drown him, have nevertheless admitted that there are other situations where the distinction is on the face of it not quite so easy to dismiss.[14] Thus they ask us to consider being faced with the conflict whether to kill one person and thereby save several other people (who need, say, that person's vital organs if they are to survive), or simply to let these people die.

Now these critics candidly admit that since it would be wrong to kill the first person even in order to save the others, we ought just to let them die. However they account for this judgment in terms of an asymmetry between the alternatives, which does not depend upon their contrasting characters as killing and letting die. For them, the morally significant difference is that the first alternative would make one person's death contribute to the well-being of all the others, whereas letting these others die would not similarly make them contribute to his well-being.[15] So, only the first alternative involves sacrificing one person for the sake of others, and hence contravenes the fundamental principle, originating with Kant but now worked into our ordinary morality, that one ought never to use one person as a 'mere means' for others' ends. But though the Kantian principle is here contravened by killing rather than by refraining from saving, a situation can easily be envisaged where one person is deliberately allowed to die in order to benefit others, and which would consequently also contravene the Kantian principle. Suppose, for instance, a beggar is deliber-

[14] See for instance Bruce Russell, 'On the relative strictness of negative and positive duties', *The American Philosophical Quarterly* (1977), pp. 87–97.

[15] Ibid. p. 95.

ately allowed to starve to death, so that his body can be used for research.

The conclusion is again, even with these apparently harder cases, that all the moral work is done by factors extrinsic to the acts/omissions distinction, which is itself apparently laid bare as a purely empirical non-moral distinction between the agent moving and not moving his body in relevant ways. Does this mean that we must after all abandon the acts/omissions doctrine, and with it any reliance that might be placed on it in the persistent vegetative state dilemma?

On the contrary, this last example should prompt us to question this whole strategy of attack. How can these critics be so sure that letting the beggar starve to death, in the above example, ought to be classified as merely refraining from saving his life, rather than as killing him? Indeed if categorizing something as an omission *necessarily* mitigates its wrongness, as the proponents of the acts/omissions distinction hold, they could simply reply that such obvious wrongs as letting someone starve to death or (as in Rachels' example) standing by and gleefully watching one's cousin drown, are just not omissions at all.

Now if Jones's doing nothing to save the drowning child could somehow be classified as an action, Rachels' example would no longer demonstrate that the acts/omissions doctrine conflicted with our ordinary moral perceptions since there would seem now to be no reason for its proponents themselves to hold back from the view that Jones was equally as iniquitous as Smith.

But critics of the doctrine will now bring into the open what has been their assumption all along, and object that to call 'doing nothing' an 'action' is to abandon what is central to the doctrine, namely the purely empirical consideration of whether an agent performs or fails to perform the relevant movements. For Bruce Russell, for instance, x kills y 'if x causes y's death by performing movements which affect y's body such that y dies as a result of these movements' whereas x lets y die 'if there are conditions (C) affecting y such that if they are not altered, y will die' and 'x fails to perform movements [in his power]... which had he performed them would have prevented y from dying, and y dies as a result of conditions (C)'.[16]

A swift glance at the literature,[17] however, suggests that it is gen-

[16] Ibid. p. 87.

[17] See again for instance Rachels, 'Active and passive euthanasia' and Russell, 'On the relative strictness' and also M. Tooley, 'An irrelevant consideration: killing versus letting die', in J. M. Fischer and M. Ravizza (eds), *Ethics, Problems and Principles* (New York: Harcourt, Brace Jovanovich, 1992), pp. 106–111.

erally the opponents of the acts/omissions doctrine who insist on the so-called empirical non-moral version of the distinction, contrasting action with *inaction*. But if the doctrine when thus interpreted turns out to be morally counter-intuitive when applied to certain key examples, its proponents are still free to try to reformulate it in a more plausible way.

(iii) A radical way with the critics: challenging their empirical interpretation

A notable attempt to loosen this empirical strait-jacket is Philippa Foot's. In her article 'Abortion and the doctrine of the double effect',[18] she proposes that the terms 'doing' and 'allowing' be substituted for 'act' and 'omission'. Her reasons are that 'omission', as ordinarily used, signifies inaction, whereas one could speak, for instance, of A 'allowing' B to die, if A is morally exculpated from murder, even when A acted to remove an obstacle to B's death, and did not not merely 'forbear to prevent' B's death.

For Foot letting someone die is not necessarily inaction, but she does not go so far as to deny that killing necessarily involves action. Some have however taken this further step and so rejected altogether the action/inaction version of the doctrine. Examples of this extreme view that doing nothing is sometimes an action are scattered across the literature, and the following is to be found in Jonathan Bennett's famous article 'Whatever the consequences':

> If on a dark night X knows that Y's next step will take him over the edge of a high cliff, and he refrains from uttering a simple word of warning because he does not care or because he wants Y dead, then it is natural to say not only that X lets Y die, but that he kills him—even if it was not X who suggested the route, removed the fence from the cliff-top, etc.[19]

What is significant here however is not simply that X is described as killing Y, even though he remains motionless, but that he is so described on the grounds, and solely on the grounds, that his conduct is in the circumstances iniquitous.

The view that our moral judgments can in this way directly determine whether a course of conduct leading to another's death is to count (in Foot's terminology) as 'doing' or merely 'allowing' has been held in a systematic way, though usually over a limited

[18] *Virtues and Vices* (Oxford: Basil Blackwell, 1981), pp. 19–32.
[19] *Analysis* (1966); reprinted in Fischer and Ravizza, *Ethics, Problems and Principles*, pp. 93–105.

range of cases, by a long and distinguished pedigree of philosophers.

St Thomas Aquinas held that 'a lack of action' counts as 'voluntary', and as such is to be classed with 'doings', and is thus blameworthy whenever a person could, and should, have acted.[20] It would seem then that according to St Thomas if one has a duty to prevent death (as typically have doctors), and when one could have prevented it, failure to prevent it is to be counted as a doing and would be, for this reason, just as iniquitous as directly killing.

The converse claim would be that even if someone's death were a direct result of our deliberate actions, we might nevertheless still be regarded as having let the person die because, and solely because, our conduct could be held to have been legitimate in the circumstances. It has recently been maintained, for example, that a doctor's active withdrawal of life-sustaining treatment from a patient counts as 'killing' just in so far as it is done *illegitimately*, that is to say without the patient's or the family's consent.[21] If the withdrawal were legitimate because it was done with the patient's and family's consent, then it would count as just letting the patient die.[22]

It is perhaps hardly surprising however that, to many, applying terms, such as 'killing' or 'letting die', directly on the basis of our moral judgments is theoretically unsatisfactory. Without any empirical or factual underpinning the moral claim that a doctor,

[20] *Summa Theologica*, XVII (Cambridge, England: Blackfriars, 1970), 1a2ae Q.6 article 3, pp. 15–16: 'To be voluntary means to spring from the will. Now one may come from another in two ways ... directly, when it proceeds from it precisely as agent, thus heating from fire ... indirectly, when it proceeds from it precisely as not acting, thus a shipwreck from loss of helm ... the result of a lack of action is not always to be brought home to the non-acting agent, but only when he could and should have acted ... now there are cases when by its resolve and action the will can intervene to break the inertia with respect to willing and acting, and sometimes ought to do so. Then it can be held responsible, for the not willing and not acting are in its charge. Thus there can be voluntariness without an act...'

[21] See A. D. Woozley, 'A duty to rescue: some thoughts on criminal liability', *Virginia Law Review*, Vol. 69 (1983), pp. 1273–1300, particularly p. 1297.

[22] Foot ('Abortion') herself seems at one point to appeal directly to a moral criterion. Thus she claims (p. 28) that for someone deliberately to allow a beggar to starve to death is the violation of a negative duty. This despite the fact that she had earlier (p. 27) observed that negative duties are typically fulfilled by refraining from the relevant actions and positive duties by performing them.

who withdraws life-sustaining treatment without a family's consent kills the patient rather than merely letting him die apparently lacks any further foundation.[23]

(iv) Infusing the empirical interpretation with moral significance

The rejection of this latter approach does not however necessarily throw us back upon a *purely* empirical and non-moral version of the doctrine, such as the action/inaction contrast, leading eventually to the undermining of any reliance that might be placed upon the doctrine in the medical situation. Fortunately, an intermediate interpretation in which empirical and moral elements are indissolubly fused has recently been expounded most notably by Warren Quinn[24] and Jeff McMahan.[25]

Although, as I now hope to show, Quinn's version of the doctrine is fatally flawed, McMahan's is sufficiently convincing for us to place some reliance upon it in the practical context. It also has the added attraction of being built around cases of the active withdrawal of life-saving aid making it highly relevant to the medical situation in which we are interested.

(a) *Quinn: tinkering with the empirical interpretation.* The difference between Quinn's and McMahan's approaches, and that of those who adhere to the purely empirical version is most tellingly revealed at the level of method. Thus Rachels refuses to allow his intuitive moral beliefs about particular cases to have any role in shaping the distinction, though these are later used to test whether the distinction has any moral validity. Quinn and McMahan on the other hand use at the start their intuitive moral beliefs about certain critical cases in order to formulate the distinction.

Quinn's reformation may seem, compared with McMahan's, to be little more than tinkering with the action/inaction version. Thus despite following Foot in calling it a doctrine of 'doing and allowing', he largely reinstates the action/inaction contrast, though with one qualification. That is to say, the active category (his 'harmful positive agency'), though remaining largely made up of actions leading to harms, is extended to include harms coming

[23] See Bennett, 'Whatever the consequences', p. 94. 'I want to see what difference there is between killing and letting die which might be a basis for a moral judgment.'

[24] 'Actions, intentions and consequences: the doctrine of doing and allowing', *The Philosophical Review*, **99** (1990), pp. 131–155. This article will henceforth be referred to as *AIC*.

[25] 'Killing, letting die, and withdrawing aid', *Ethics*, **103** (January 1993), pp. 250–279. This article will henceforth be referred to as *KLW*.

from 'a special kind of inaction'. What Quinn means here is perhaps best understood from one of his examples:

> We are off by special train to save five who are in imminent danger of death. Every second counts. You have just taken over from the driver, who has left the locomotive to attend to something. Since the train is on automatic control you need do nothing to keep it going. But you can stop it by putting on the brakes. You suddenly see someone trapped ahead on the track. Unless you act he will be killed. But if you do stop, and then free the man, the rescue mission will be aborted. So you let the train continue. (*AIC*, p. 298)

Now since, according to Quinn (*AIC*, p. 299), just letting the train continue is to 'make the [morally] wrong choice', your conduct ought to count, to use his expression as 'harmful positive agency', rather than 'harmful negative agency', even though you did nothing. But then you kill the man on the track rather than just letting him die. Quinn justifies speaking in this way by pointing out that you not only had control over whether or not the train ploughed into the man, but also intended it: 'the train kills the man because of your intention that it continue forward'.

He further clarifies his point by slightly altering the scenario so that now you still have the relevant control, but not the relevant intention:

> Suppose you are on a train on which there has just been an explosion. You can stop the train, but that is a complicated business that would take time. So you set it on automatic forward and rush back to the five badly wounded passengers. While attending to them, you learn that a man is trapped far ahead on the track. (*AIC*, p. 299)

Quinn believes (*AIC*, p. 300) that it is now permissible for you to choose to stay with the wounded passengers, and help them, rather than leaving them to stop the train and save the man because here you 'merely tolerate the action of the train that causes the man's death'. By contrast in the first scenario 'your choice to let the train continue forward is strategic and deliberate' and 'since you clearly *would* have it continue for the sake of the five there is a sense in which, by deliberately not stopping it, you do have it continue'.

Quinn provides (*AIC*, p. 306) a moral underpinning for the doctrine in terms of the greater stringency of negative over positive rights.[26] Thus the man on the track has a greater right not to be killed by the train than do the others to be saved.

This precedence of negative over positive rights is itself subsumed (*AIC*, pp. 308–312) under a still more general principle that people are 'ends in themselves', and as such it is morally imperative that they have a 'say' in what is done to them. But if they only have a 'say' when what they want would not compromise the collective interests of others, then this would really be, as Quinn puts it, no say at all. Indeed it is compatible with one's being allowed to do what harm one likes to another person so long as it is collectively best.[27]

But there are a number of problems with this account. If for Quinn one does harm when one sacrifices one person's interests to others' then he has to show why, with the exception of situations such as his train set to non-stop, doing such harm necessarily involves *action*, as he claims it does. Our earlier example of not giving food to a starving beggar in order that his body could be used for research purposes seemed to show, quite contrarily, that one could sacrifice people for others' ends, and hence do harm to them in Quinn's sense, whilst remaining inactive. We seem to have two criteria here pulling against each other: action and inaction contrasted with whether individuals are being treated as mere means for others' ends or as ends in themselves.

A connected difficulty arises from Quinn's exclusive concentra-

---

[26] Quinn points to Foot as having first suggested such an underpinning for the doctrine. Foot's underpinning however was in terms of the distinction between positive and negative *duties*, rather than positive and negative rights. For her, only negative *duties* are correlated with rights, since being owed to each and everyone, their fulfilment is a matter of justice. Positive duties belong merely within the sphere of beneficence: their fulfilment, not being owed to each and everyone, may not be claimed by each and everyone as his right.

[27] Both Quinn's and McMahan's accounts of the doctrine of acts/omissions have the singular attraction of explicitly attempting to link it with the general moral system of which they see it as an intrinsic part. Only utilitarians have so far spelt out how their approach to the doctrine (namely their opposition to it) is linked with the larger moral system that they endorse. For Quinn, the doctrine reflects a moral scheme in which the fundamental concern is rights, and in particular, the right to live one's life as one chooses. It further turns out that these rights, rather like Nozickian rights, are protections for the individual against unbridled utilitarianism. I am not convinced however that our ordinary morality, which Quinn so frequently appeals to when trying to determine our moral judgment of particular types of conduct, really is so systematically right-based.

tion on what we earlier called 'conflict situations', since these are necessarily situations where *different* peoples' interests conflict. Since voluntary active euthanasia is aimed at benefiting only the person who requested it, it would not seem to count as a 'doing', a killing, by Quinn's *moral* criterion, yet would seem to count as such by his *empirical* criterion.

This failure to fully infuse the empirical element with moral significance leaves a worrying lack of integration at the core of his account. This springs as much from his initial narrow characterization (*AIC*, p. 287) of the doctrine as enshrining the prohibition against harming someone *to help others*, as from his clinging to the action/inaction distinction.

(b) *McMahan: why some withdrawals are killing and others merely letting die.* McMahan starts out far more sceptically than Quinn. Choosing to examine the doctrine through the killing/letting die distinction he observes that if there is an underlying factual basis for the evident moral distinction here it is certainly obscure. For the reasons given above, it cannot simply be a question of whether the relevant bodily movements are or are not made. That is to say, it cannot boil down to the distinction between action and *inaction*. Thus he takes a series of critical examples in which someone dies as a result of another person's conduct, and appealing jointly to his ordinary moral intuitions (how heinous is that?) and ordinary usage (would we typically call that 'killing'?), he sorts the cases of killing into one group and the cases of letting die into another group. He then looks for a common feature, having both an empirical and a moral dimension, present in all the cases of killing and absent in all the cases of letting die, which might account for the different moral judgments we make in the two groups.

His first example (*KLD*, pp. 251–252) is of a swimmer who attempting in a dangerous sea to save a drowning man, but finding himself in difficulties, pushes the man off with the result that though the man drowns, the swimmer survives. Appealing both to ordinary moral intuition and ordinary usage McMahan claims that the swimmer could not possibly be accused of killing the drowning man.

He then, however, produces a further example (*KLD*, p. 254) meant to show that we should not therefore conclude that, as Foot would have it, whenever someone merely removes a barrier which is keeping death at bay he is letting the person die rather than killing him. The interloper in a hospital who comes upon his enemy on a life support machine when no-one else is around, and seeing this as an opportunity to get rid of him, quickly detaches him from it, surely kills the patient.

McMahan's interim conclusion (*KLD*, pp. 255–256) is that what distinguishes the cases of killing from the cases of letting die is whether or not the barrier to death that is removed is one that the individual himself has provided: withdrawing aid in such circumstances simply nullifies the initial intervention. On the other hand withdrawing aid which the individual did not himself provide can be thought of as creating a new threat.

This account is refined with a further example (*KLD*, p. 256). Suppose a workman, who had previously sealed a cracked pipe from which had been pouring poisonous chemicals, now returns and deliberately removes the sealant, releasing the poisonous chemicals once more with fatal consequences for the inhabitants of the nearby houses.

According to McMahan the workman may be described as killing the inhabitants since the pipe was in working order and no longer required repairing, and he concludes that only if the aid provided by an individual requires more effort from him to sustain it, that is to say the initial aid was not 'self-sustaining', will removing it count as merely letting the other person die.[28] The 'primary cause' as he calls it is not in the circumstances the individual's intervention at all, but antecedent conditions. If on the other hand the aid is already self-sustaining and like the workman, an individual removes it, then 'the primary cause' is the individual's intervention, and he is then considered to be actively killing rather than merely allowing death to occur.

For McMahan, the moral significance of whether active intervention is considered to be a 'primary cause' of death is to be found in our ordinary morality, which defines not only our core notion of causation as an active force but also, as he puts it, 'the form and degree of an agent's moral responsibility for a person's death'.[29]

[28] In order to keep McMahan's account simple, I have left out one further condition that he stipulates for the withdrawal of life-saving aid to count as killing, namely that it should be 'operative' (*KLD*, p. 261).

[29] McMahan's view of the role of causation in the acts/omissions distinction can be illuminatingly contrasted with that of Quinn's. For Quinn, causation is irrelevant to the distinction since according to him even a person who is inactive can nevertheless be said to cause, or to partially cause, death. For McMahan, on the other hand, not only is being the 'primary cause' (in the sense defined) of another's death necessary and sufficient for killing, but causation is necessarily 'active'. Moreover he sees the key concern of ordinary morality as that of attributing moral responsibility to an agent as an active cause, rather than, as that of (*pace* Quinn) ensuring that an individual's right to live his life as he chooses is respected. Although I believe that McMahan's is more plausible as an

This view, that active intervention is necessary for killing, has as a consequence that inaction can never be killing: McMahan can only deal with Rachels's example of Jones standing by and watching his cousin drown as, not a killing, but as nevertheless morally wicked.

On the other hand, as we have seen, though killing, for McMahan, necessarily involves active intervention, such interventions leading to death are not necessarily killings, but may be merely letting die. Hence his version of the acts/omissions doctrine allows us to dismiss one familiar objection to counting the doctor's withdrawal of food and water from the patient in PVS a 'letting die', namely that it is an active intervention.[30]

With this apparently promising version of the doctrine in hand, conveniently focused on cases of the active withdrawal of life-saving aid, it might seem that all that is left for us to do is to apply it, almost mechanically, to the medical moral dilemma, as though the philosopher had nothing to learn from the practical situation which his theories had not already envisaged.

In subjecting the notion of an omission to this prolonged philosophical analysis, I have of course already gone a good deal further than typical 'medical ethicists' who simply ignore the deep philosophical roots of the concepts to which they make appeal, as irrelevant for decision-makers at the 'sharp end'.

I will now go beyond a general philosophical discussion of the acts/omissions doctrine, by applying it to the PVS dilemma in order to bring out the inadequacies of the version just expounded. I hope to show that McMahan's use of hypothetical and to some extent fanciful examples, unchecked by the exigencies of an actual situation, involving special duties and responsibilities, such as the doctor's, has led him systematically to overlook a feature which often explains our conviction that certain kinds of conduct ought to count as killing, and which must be re-incorporated at the theoretical level. I hope thus to show that it is not merely that practical decision-makers and legislators have something to learn from philosophers, but that philosophers in turn too are more likely to produce good or valid theories, if they would only study actual dilemmas.

---

[30] See Lord Goff, H.L. 396D.

---

account of ordinary morality than is Quinn's, I do not think that causation, in this active sense, is the only criterion used in ordinary morality for determining whether an agent should be held to have killed someone. The intention to bring about death is also crucial; see below.

2. Doctors, Patients and the Persistent Vegetative State

(i) Applying McMahan's criteria: causation and intention inside and outside the medical situation

Since the doctor, like the swimmer, is prima facie himself providing the life-sustaining aid, and since further effort from him is required, it might at first sight seem that, like the swimmer, he does not create a new threat in withdrawing aid. He cannot then be regarded as, in McMahan's sense, the primary cause of the patient's death, and so could not be said to kill the patient but merely to let him die. But this is too swift.

The doctor himself is a dispensable element in a hospital network of aid to the patient. Were he to stop giving aid, his place would normally be taken by someone else, and so the aid is self-sustaining in the sense that it is not dependent on any particular individual acting on his own, but rather as a member of a team whose job it is to provide aid.

Now if the doctor as leader of the team decides to withdraw the aid, and agrees that no-one else from the team shall be brought in to sustain it, then the doctor has clearly intervened in the self-sustaining mechanism, and so created a new threat to the patient's life. It must surely then follow, on McMahan's analysis, that the doctor is not only the primary cause of the patient's death, but kills the patient. But this also is too swift.

If whenever a doctor could be described as the primary cause of his patient's death, he also 'killed' his patient, it would seem that doctors 'kill' their patients in a number of circumstances where even the most conservative moralists would hold that what they do is not only permissible but even perhaps required.

McMahan's account of the killing/letting die distinction is perhaps plausible outside the medical context because it is unlikely (though not impossible) that anyone would deliberately withdraw life-saving aid from another person so as to create a new threat to his life, unless his *intention* was also to bring about this other person's death. In the everyday situation, the main reason that it is so important to establish that the maintenance of aid would require further costly personal effort from the rescuer is that it helps rule out the possibility that in deliberately withdrawing this aid, his intention is expressly to bring about the victim's death. The swimmer for instance is not prepared to continue to jeopardize his own life, or is perhaps simply fatigued, and this accounts for his pushing the man away from him, not an intention to drown him.[31]

[31] This is not to deny that self-preservation or fatigue also serve, after it has been established that he had no intention to bring about death, to exonerate the swimmer from not continuing to give aid.

But once we appreciate the importance of ruling out the intention to bring about death in such circumstances, we might begin to suspect that it is this very intention which makes the interloper's withdrawal of life-saving aid from his enemy, unlike the swimmer's, so decisively a case of killing rather than merely letting someone die.

In the medical context however, causing and intending death are not, typically, so closely linked. This is because the doctor may have aims in withdrawing (or indeed in withholding) life sustaining treatment *other* than that of bringing about the patient's death, and just for this reason, though he causes the patient's death, it may plausibly be denied that he kills her.

A doctor might withdraw life-sustaining treatment at a patient's own request, simply in order to respect her right to self-determination. Here the doctor could hardly be said to have wanted or intended her death, and may well have argued passionately with her that he should not withdraw the treatment since this would lead to her death.[32]

Or again, a doctor with a patient suffering from terminal cancer might withdraw a painful treatment he is administering, which is merely abating another condition, and which if abated would probably lengthen his life by a few days.[33] Here although death comes about sooner than it might otherwise have done, the doctor's intention in withdrawing the painful treatment is not, it may be said, to hasten death, but merely to spare the patient further painful and pointless suffering, of which this hastening of death is a foreseen but unintended side-effect.[34]

In both these cases a further question is whether the doctor's conduct is morally justifiable. If the doctor were, in McMahan's terms, the primary cause of the patient's death, we could not, if we continue to follow McMahan, appeal to the acts/omissions doctrine. But though the doctor does not merely let the patient die, his conduct, I have suggested, cannot, either, be proscribed as 'killing', once this term requires the deliberate intention to bring about death.

Since however in both cases death is an unintended side-effect

---

[32] See Lord Goff, (*H.L.*, 367E)

[33] I here draw on a situation described by Lord Goff at 370C.

[34] One other legal possibility is suggested by Lord Browne-Wilkinson at 385–386: since legally a doctor may only continue with an intrusive life support system if it is in the best interests of the patient, if a responsible body of medical opinion should judge it not in the patient's best interests, the doctor must withdraw life-support if he is not to commit battery.

of the intentional pursuit of another good, it would seem that the doctor's conduct might be justified by appeal to another major doctrine of ordinary morality, the doctrine of double effect. Whether the doctor's conduct can be successfully justified in these terms will of course depend upon whether the intended good really does outweigh the unintended evil of the patients' deaths.[35]

Returning now to the PVS patient, it would thus seem that to show that the doctor withdraws food and water with an intention other than to bring about the patient's death would serve a double function. It would exclude the intention to bring about death, and hence deflect the charge of killing, and yet also provide a putative justification for nevertheless deliberately causing the patient's death.

But the doctor could hardly be described as withdrawing food and water from a PVS patient in order, as in the first case, to respect her right to refuse treatment. Such a patient, who has not previously made a living will (which is usually the case with young accident victims) neither has made, nor could make, a request that their treatment be terminated.

Nor can it seriously be suggested that the doctor wants to spare his PVS patient further pointless suffering, since she is already insensate. But maintenance of life support for these patients is not pointless since it could keep them alive for many years. Nor does it seem as if there is any other intention which would rule out a deliberate intention to end their lives by withdrawing treatment.

---

[35] McMahan (*KLD*, p. 273) himself points out that 'factors affect the moral status of a course of conduct that has lethal consequences other than the distinction between killing and letting die... the most commonly noted... is whether a person's death is an unintended effect of the agent's action'. But, as the medical examples show, discovering that death was not intended does not just affect our view of the 'moral status' of a course of conduct already established as a 'killing', rather it affects whether we *call* it 'killing'. Though, therefore, I largely agree with McMahan's criteria for 'letting die'—the relevant course of conduct being a candidate for certain characteristic types of justification, such as that the personal cost of acting would have been too great for the agent to bear, or that it involved contravening some other more stringent prohibition—I disagree with his criteria for 'killing'. McMahan could distinguish between two types of killing, only one of which would be absolutely ruled out, namely where the intention had been to bring about death. But then it would have to be made clear that even conduct falling on the 'killing' side of the divide might conceivably turn out to be morally justified, namely where death was an unintended side effect of the pursuit of some other aim. These further determinations of moral status would now however fall outside the ambit of the acts/omissions doctrine, requiring appeal to double effect.

But though the deliberate intention to terminate these lives by withdrawing treatment cannot be ruled out, on what grounds, if any, might we *positively impute* such an intention?[36]

(ii) Does the doctor intend to bring about death?

Consider a situation where doctors decide to withhold life-saving treatment from a severely handicapped neonate in circumstances where the child's life, if saved, would be irredeemably racked by pain and agony.[37] Here it might seem that the withholding of treatment from the neonate is motivated by a desire to release it from a future existence of suffering by terminating its life.

However it might still plausibly be denied that the withholding of life-saving treatment, even in what are patently the child's best interests, entails an active intention to bring about the child's death. Appeal would typically be made to what is sometimes called the 'open door' argument: withholding treatment leaves the possibility that there might be a sudden and unexpected remission in the neonate's condition—which sometimes happens—in contradistinction to using a lethal injection to terminate life.

Thus in situations where life-saving treatment is deliberately withheld from severely handicapped babies, one of the criteria for the attribution of an intention to bring about death is typically present, viz. a motive for bringing that death about, but the other, and possibly more compelling, criterion for this attribution, viz. that all avenues to survival are deliberately closed, is absent.

With the PVS situation however it is the more compelling criterion that is present and the less compelling criterion that is absent. That is to say, on the one hand there is no motive for terminating life similar to that in the case of severely handicapped babies. Since the PVS patient is insensate, and so presumably not suffering, it cannot be claimed that terminating his life would be in his best interests. But on the other hand, the powerful 'open door' argument, which could have been used to rebut the suggestion that it is intended that the patient die, is inapplicable here, since it is assumed that there is no possibility of his survival, once food has been withdrawn.[38]

[36] Lord Browne-Wilkinson at 383E (*H.L.*) holds that such an intention must be imputed to the doctor in these situations. R. Gillon however denies this in "Patients in the persistent vegetative state: a response to Dr Andrews", *BMJ*, (Vol. 306, 12th June 1993), pp. 1602–1603.

[37] J. Lorber, 'Ethical problems in the management of myelomeningocele and hydrocephalus', *Journal of Royal College of Physicians*, Vol. 10, No. 1 (Oct. 1975), pp. 47—60.

[38] H.L., at 372G per Lord Goff.

Now given that, as we have seen, there could in this situation be no intention in withdrawing treatment other than bringing about the patient's death (he is insensate, he has no wishes that must be respected etc.), and given that the 'open door' argument is not available here to combat the claim that this is in fact the intention, we are I believe justified in asserting both that the doctor intends to bring about the patient's death and that he not only causes her death but kills her.

If we are however at all impressed by the fact that the withdrawal of life-sustaining treatment is *not* the giving of a lethal injection, and hence is, in some sense or other, an 'omission', we may prefer to describe it as *morally equivalent* to, rather than actually, killing. But its status as an omission will now fail to have any moral relevance at all: only the wraith of the doctrine remains.

Postscript: 'new medicine [requires] a new ethics' (Lord Lowry[39])

At the start, I stressed the limited nature of the conclusion I hoped to reach: that the withdrawal of food and water from a patient in PVS was killing. It follows that our ordinary morality, the contours of which I hope have also become clearer, must proscribe it, with the dire practical consequence that hospital beds would increasingly be filled with such insensate patients.

I offer some final thoughts on the prospects for a 'new ethics', as Lord Lowry called it which, though permitting deliberate killing, might be able to fill the gap that, as Lord Lowry observed, Lord Mustill had pointed to between 'old law and new medicine'.

Some moral philosophers have gone so far as to hold the view that rights must protect, and only protect, individuals who are capable of making autonomous choices. On such a view, any right can be waived as well as exercised. Thus a patient waives her right to bodily inviolability in consenting to invasive treatment, but exercises the right when she refuses such a treatment. Furthermore they regard rights as morally fundamental in that there can be no serious wrongs that are not violations of the rights of autonomous persons.

There are two important consequences of these views. Firstly the characterization of rights as waivable opens the way for these moral philosophers to argue that it is possible for an autonomous person to waive not just her right to bodily inviolability or to freedom from interference (as in the above example) but also her right to life, that is to say both her right not to be deliberately killed but

---

[39] *H.L.*, 379.

also any right she might have to be saved. Thus on this view voluntary euthanasia is permissible.

Secondly in such a 'right-based' morality, since they are not autonomous, the senile, foetuses and the insensate would have no rights—neither the right to life nor the right to die nor any other right—and hence would not have the protection that rights afford.[40] Our attitude toward them, on this view, should be rather like our attitude to animals. We might have a duty not to cause them suffering and hence, if we wished to kill them, we would owe them a humane death, though they would have no *right* either to a humane or dignified death.

In this extreme type of morality the killing of a PVS patient would be morally permissible, but at the expense of morally permitting the killing of many other human beings, who were not thought worth protection. The difficulty is to ameliorate this harsh morality, which not only fails to draw a line between the insensate and those who are less incapacitated, and who by the standards of our ordinary morality we regard as worthy of protection, but also does not even require a justification for their killing.

One solution is perhaps to move toward a more liberal form of our ordinary morality which permits certain qualifications of the absolute prohibition on killing innocent human life. One justification both of abortion and of embryo research is that foetuses and embryos have only partial moral status which can be outweighed by the interests of others with full moral status. Perhaps the misery experienced by the family of a patient in PVS might be regarded, on this view, as outweighing any duties we might owe to the patient himself. If we followed the logic of the argument concerning abortion, we would then be prepared to kill such a patient.

However with embryos and foetuses it is simply because they are at a primitive stage of development, and not because of damaged capacities, that we are permitted to accord them this low moral status. If however we were to accord an equally low moral status to adult patients, such as those in PVS, this would now depend upon a 'quality of life judgment'. That is to say, if we were to allow the interests of the family of a patient in PVS to outweigh the patient's right to life, this would be because we assented to the proposition that, because of incapacity or infirmity, one life may be intrinsically less worthwhile than another, and may hence be sacrificed for the good of others.[41]

[40] Lord Browne-Wilkinson refers to such a view at 381C–D.
[41] See M. Warnock, *A Question of Life: The Warnock Report on Human fertilization and Embryology* (Oxford: Blackwell, 1985), 11.15, p. 62.

Technology and Culture in a Developing Country

KWAME GYEKYE*

Even though the subject of my paper is 'Technology and Culture in a Developing Country', it seems appropriate to preface it by examining science itself in the cultural traditions of a developing country, such as Ghana, in view of the fact that the lack of technological advancement, or the ossified state in which the techniques of production found themselves, in the traditional setting of Africa and, in many ways, even in modern Africa, is certainly attributable to the incomprehensible inattention to the search for scientific principles by the traditional technologists. I begin therefore with observations on how science and knowledge fared in the traditional culture of a developing country.

Science and Our Culture

In a previous publication I pointed out—indeed I stressed—the empirical orientation of African thought: maintaining that African proverbs, for instance, a number of which bear some philosophical content, addressed—or resulted from reflections on—specific situations, events or experiences in the lives of the people, and that even such a metaphysical concept as destiny (or fate) was reached inductively, experience being the basis of the reasoning that led to it.[1] Observation and experience constituted a great part of the sources of knowledge in African traditions.[2] The empirical basis of

* This paper was written when the author was a Fellow at the Woodrow Wilson International Center for Scholars, Smithsonian Institution, Washington, D.C., during the 1993/94 academic year.
    [1] Kwame Gyekye, *An Essay on African Philosophical Thought* (New York: Cambridge University Press, 1987), pp. 16–18 and 106–107.
    [2] It is instructive to note that the Ewe word for 'knowledge' is *nunya*, a word which actually means 'thing observed'. This clearly means that observation or experience was regarded as the source of knowledge in Ewe thought: see N. K. Dzobo, 'Knowledge and truth: Ewe and Akan Conceptions', in Kwame Gyekye and Kwasi Wiredu (eds), *Person and Community: Ghanaian Philosophical Studies* (Washington D.C.: The Council for Research in Values and Philosophy, 1992), pp. 74ff. The empirical character of African thought generally can most probably not be doubted.

knowledge had immediate practical results in such areas as agriculture and herbal medicine: our ancestors, whose main occupation was farming, knew of the system of rotation of crops; they knew when to allow a piece of land to lie fallow for a while; they had some knowledge of the technology of food processing and preservation; and there is a great deal of evidence about their knowledge of the medicinal potencies of herbs and plants—the main source of their health care delivery system long before the introduction of Western medicine. (Even today, there are countless testimonies of people who have received cures from 'traditional' healers where the application of Western therapeutics could not cope.)

It has been asserted by several scholars that African life in the traditional setting is intensely religious or spiritual. Mbiti opined that "Africans are notoriously religious, and each people has its own religious system with a set of beliefs and practices. Religion permeates into all the departments of life so fully that it is not easy or possible to isolate it.'[3] According to him, "in traditional life there are no atheists".[4] Busia observed that Africa's cultural heritage 'is intensely and pervasively religious',[5] and that 'in traditional African communities, it was not possible to distinguish between religious and nonreligious areas of life. All life was religious'.[6] Many colonial administrators in Africa used to refer to Africans, according to Parrinder, as 'this incurably religious people'.[7] Yet, despite the alleged religiosity of the African cultural heritage, the empirical orientation or approach to most of their enterprises was very much to the fore. I strongly suspect that even the African knowledge of God in the traditional setting was, in the context of a nonrevealed religion of traditional Africa, empirically reached.

Now, one would have thought that such a characteristically empirical, epistemic outlook would naturally lead to a profound and extensive interest in science as a theory, that is, in the acquisition of theoretical knowledge of nature, beyond the practical knowledge which they seem to have had of it, albeit not in a highly developed form, and which they utilized to their benefit. But, surprisingly, there is no evidence that such an empirical orientation of thought in traditional African culture led to the creation of the sci-

[3] John S. Mbiti, *African Religions and Philosophy* (New York: Doubleday and Company, 1970), p. 1.

[4] Ibid. p. 38.

[5] K.A. Busia, *Africa in Search of Democracy* (New York: Praeger, 1967), p. 1.

[6] Ibid. p. 7.

[7] G. Parrinder, *African Traditional Religion* (New York: Harper & Row, 1962), p. 9.

entific outlook or a deep scientific understanding of nature. It is possible, arguably, to credit people who practiced crafts and pursued such activities as food preservation, food fermentation, herbal therapeutics, etc. (see next section) with some amount of scientific knowledge; after all, the traditional technologies, one would assume, must have had some basis in science. Yet, it does not appear that their practical knowledge of crafts or forms of technologies led to any deep scientific understanding or analysis of the enterprises they were engaged in. Observations made by them may have led to interesting facts about the workings of nature; but those facts needed to be given elaborate and coherent theoretical explanations. Science requires explanations that are generalizable, facts that are disciplined by experiments, and experiments that are repeatable and verifiable elsewhere. But the inability (or, is it lack of interest?) of the users of our culture to engage in sustained investigations and to provide intelligible scientific explanations or analyses of their own observations and experiences stunted the growth of science.

Science begins not only in sustained observations and investigations into natural phenomena, but also in the ascription of causal explanations or analyses to those phenomena. The notion of causality is of course crucial to the pursuit of science. Our cultures appreciated the notion of causality very well. But, for a reason which must be linked to the alleged intense religiosity of the cultures, causality was generally understood in terms of spirit, of mystical power. The consequence of this was that purely scientific or empirical causal explanations, of which the users of our culture were somehow aware, were often not regarded as profound enough to offer complete satisfaction. This led them to give up, but too soon, on the search for empirical causal explanations, even of causal relations between natural phenomena or events, and resort to supernatural causation.

Empirical causation, which asks what and how questions, too quickly gave way to agentive causation which asks who and why questions. Agentive causation led to the postulation of spirits or mystical powers as causal agents; so that a particular metaphysic was at the basis of this sort of agentive causation. According to Mbiti, 'The physical and spiritual are but two dimensions of one and the same universe. These dimensions dove-tail into each other to the extent that at times and in places one is apparently more real than, but not exclusive of, the other.'[8] It is the lack of distinction between the purely material (natural) and the immaterial (super-

[8] Mbiti, *African Religions and Philosophy*, p. 74.

natural, spiritual) that led to the postulation of agentive causation in all matters. For, in a conception of a hierarchy of causes, it was easy to identify the spiritual as the agent that causes changes in relations even among empirical phenomena. In view of the critical importance of causality to the development of the science of nature, a culture that was obsessed with supernaturalistic or mystical causal explanations would hardly develop the scientific attitude in the users of that culture, and would, consequently, not attain knowledge of the external world that can empirically be ascertained by others, including future generations.

Yet, the alleged intense religiosity of the African cultural heritage need not have hindered interest in science, that is, in scientific investigations both for their own sake and as sure foundations for the development of technology. Religion and science, even though they perceive reality differently, need not, nevertheless, be incompatible. Thus it is possible for religious persons to acquire scientific knowledge and outlook. But to be able to do so most satisfactorily, one should be able to separate the two, based on the conviction that purely scientific knowledge and understanding of the external world would not detract from one's faith in an ultimate being. A culture may be a religious culture, even an intensely religious culture at that; but, in view of the tremendous importance of science for the progress of many other aspects of the culture, it should be able to render unto Caesar what is Caesar's and unto God what is God's ('Caesar' here referring to the pursuit of the knowledge of the natural world). The inability of our traditional cultures to separate religion from science, as well as the African conception of nature as essentially animated or spirit-filled (leading to the belief that natural objects contained mystical powers to be feared or kept at bay or, when convenient, to be exploited for man's immediate material benefit), was the ground of the agentive causal explanations enamoured of the users of our cultures in the traditional setting. Science, as already stated, is based on a profound understanding and exploitation of the important notion of causality, that is, on a deep appreciation of the causal interactions between natural phenomena. But where this is enmeshed with—made inextricable from—supernaturalistic moulds and orientations, it, as a purely empirical pursuit, hardly makes progress.

Also, religion, even if it is pursued by a whole society or generation, is still a highly subjective cognitive activity, in that its postulates and conclusions are not immediately accessible to the objective scrutiny or verification by others outside it. Science, on the contrary, is manifestly an objective, impartial enterprise whose conclusions are open to scrutiny by others at any time or place, a

124

scrutiny that may lead to the rejection or amendment or confirmation of those conclusions. Now, the mesh in which both religion and science (or, rather the pursuit of science) found themselves in African traditional cultures, made the relevant objective approach to scientific investigations into nature well-nigh impossible. Moreover, in consequence of this mesh, what could have become scientific knowledge accessible to all others became an esoteric knowledge, a specialized knowledge, accessible only to initiates probably under an oath of secrecy administered by priests and priestesses, traditionally acknowledged as the custodians of the verities and secrets of nature. These custodians, it was, who 'knew', and were often consulted on, the causes of frequent low crop yield, lack of adequate rainfall over a long period of time, the occurrence of bush fires, and so on. Knowledge-claims about the operations of nature became not only esoteric but also, if for that reason, personal rather than exoteric and impersonal. This pre-empted the participative nature of the search for deep and extensive knowledge of the natural world; for, others would not have access to, let alone participate in, the type of knowledge that is regarded as personal and arcane.

Knowledge of the potencies of herbs and other medicinal plants was in the traditional setting probably the most secretive of all. Even if the claims made by African medicine men and women of having discovered cures for deadly diseases could be substantiated scientifically, those claims cannot be pursued for verification since their knowledge-claims were esoteric and personal. The desire to make knowledge of the external world personal has been the characteristic attitude of our traditional healers who claim to possess knowledge of medicinal plants, claims at least some of which can be scientifically investigated. In the past, all such possibly credible claims to knowledge of medicinal plants just evaporated on the death of the traditional healer or priest. And science, including the science of medicine, stagnated.

I think that the personalization of the knowledge of the external world is attributable to the mode of acquiring that knowledge: that mode was simply not based on experiment. And, in the circumstance, the only way one could come by one's knowledge of, say, herbal therapeutics, was most probably through mystical or magical means, a means not subject to public or objective scrutiny and analysis.

The lack of the appropriate attitude to sustained scientific probing, required for both vertical and horizontal advancement of knowledge, appears to have been a characteristic of our African cultural past. One need not have to put this want of the appropriate scientific attitude to the lack of the capacity for science. And I, on my part, would like to make a distinction here: between the having of the intellectual capacity on the one hand, and the having of the proclivity or impulse to exercise that capacity on a sustained basis that would yield appreciable results on the other hand. The impulse for sustained scientific or intellectual probing does not appear to have been nurtured and promoted by our traditional cultures.

It appears in fact that the traditional cultures rather throttled the impulse towards sustained and profound inquiry for reasons that are not fully known or intelligible. One reason, however, may be extracted from the Akan (Ghanaian) maxim, literally translated as: 'if you insist on probing deeply into the eye sockets of a dead person, you see a ghost'.[9] The translation is of course not clear enough. But what the maxim is saying is that curiosity or deep probing may lead to dreadful consequences (the ghost is something of which most people are apprehensive). The maxim, as Laing also saw, stunts the 'development of the spirit of inquiry, exploration and adventure'.[10] The attitude sanctioned by the maxim would, as Laing pointed out, be 'inimical to science';[11] but not only to science, but, I might add, to all kinds of knowledge. My colleague, Opoku,[12] however, explained to me in a conversation that the intention of the maxim is to put an end to a protracted dispute which might tear a family or lineage apart: a dispute that has been settled, in other words, should not be resuscitated, for the consequences of the resuscitation would not be good for the solidarity of the family. Thus, Opoku would deny that this maxim is to be interpreted as damaging to intellectual or scientific probing. In response to Opoku's interpretation, one would like to raise the following questions: why should further evidence not be looked for if it would indeed help settle the matter more satisfactorily? Why should further investigation be stopped if it would unravel fresh evidence and lead to what was not previously known? To end a dispute prematurely for the sake of family solidarity to the dissatisfaction of some members of the family certainly does violence to the pursuit of moral or legal knowledge. So, whether in the area of legal, moral or scientific knowledge, it seems to me that the maxim places a damper on the impulse or proclivity to deep probing, to the pursuit of further knowledge.

[9] The Akan version is: *we feefee efun n'aniwa ase a, wohu saman.*

[10] E. Laing, *Science and Society in Ghana*, The J. B. Danquah Memorial Lectures (Ghana Academy of Arts and Sciences, 1990), p. 21.

[11] Ibid. p. 21.

[12] Kofi Asare Opoku of the Institute of African Studies, University of Ghana, Legon.

The general attitude of the users of the African traditional cultures expressed in oft-used statements as 'this is what the ancestors said', 'this is what the ancestors did' and similar references to what are regarded as the ancestral habits or modes of thought and action, may be put down to the inexplicable reluctance—or lack of the impulse—to pursue sustained inquiries into the pristine ideas and values of the culture. It is this kind of mind-set, one might add, which often makes the elderly people even in our contemporary (African) societies try to hush and stop children with inquisitive minds from persistently asking certain kinds of questions and, thus, from pursuing intellectual exploration on their own. (I do not have the space to provide evidence to show that our forefathers did not expect later generations to regard their modes of thought and action as sacrosanct and unalterable, and to think and act in the same way they did. So that, if later generations, i.e. their descendants, failed to make changes, amendments or refinements such as may be required by their own times and situations, that would have to be put down to the intellectual indolence or shallowness of the descendants.)

Finally let me say this: The pursuit of science—the cultivation of rational or theoretical knowledge of the natural world—seems to presuppose an intense desire, at least initially, for knowledge for its own sake, not for the sake of some immediate practical results. It appears that our cultures had very little if any conception of knowledge for its own sake. It had a conception of knowledge that was practically oriented. Such an epistemic conception seems to have had a parallel in the African conception of art. For, it has been said by several scholars[13] that art was conceived in the African traditional setting in functional (or, teleological) terms, that the African aesthetic sense did not find the concept of 'art for art's sake' hospitable. Even though I think that the purely aesthetic element of art was not lost sight of, this element does not appear to have been stressed in African art appreciation, as was the functional conception. This practical or functional conception of art, which dwarfed a conception of art for art's sake, must have infected the African conception of knowledge, resulting in the lack of interest in the acquisition of knowledge, including scientific knowledge, *for its own sake*.

[13] Robert W. July, for example, says: 'Art for art's sake had no place in traditional African society' and that it was 'essentially functional'. See his *An African Voice: The Role of The Humanities in African Independence* (Durham: Duke University Press, 1987), p. 49; also Claude Wauthier, *The Literature and Thought of Modern Africa* (London: Heinemann, 1978), pp. 173–174.

It is clear from the foregoing discussion of the attitude of our indigenous African cultures to science that: (i) the cultures did not have a commitment, however spasmodic, to the advancement of the scientific knowledge of the natural world; (ii) they made no attempts, however feeble, to investigate the scientific theories underpinning the technologies they developed, as I will point out in some detail below; (iii) the disposition to pursue sustained inquiries into many areas of their life and thought does not seem to have been fostered by our African cultures; and (iv) the successive generations of the participants in the culture could not, consequent upon (iii), augment the compendium of knowledge that they had inherited from their forefathers, but rather gleefully felt satisfied with it, making it into a hallowed or mummified basis of their own thought and action. In our contemporary world, when sustainable development, a great aspect of which is concerned with the enhancement of the material well-being of human beings, depends on the intelligent and efficient exploitation of the resources of nature—an exploitation that can be effected only through science and its progeny technology, the need to cultivate the appropriate scientific attitudes is an imperative.

Contemporary African culture will have to come to terms with the contemporary scientific attitudes and approaches to looking at things in Africa's own environment, attitudes and approaches that have been adopted in the wake of the contact with the Western cultural traditions. The governments of African nations have for decades been insisting on the cultivation of science in the schools and universities as an unavoidable basis for technological, and hence industrial, advancement. More places and facilities are made available for those students who are interested in the pursuit of science. Yet, *mirabile dictu*, very many more students register for courses in the humanities and the social sciences than in the mathematical and natural sciences. Has the traditional culture anything to do with this lack of real or adequate or sustained interest in the natural sciences, or not?

Technology and our Culture

Like science, technology—which is the application of knowledge or discovery to practical use—is also a feature or product of culture. It develops in the cultural milieu of a people and its career or future is also determined by the characteristics of the culture. Technology is an enterprise that can be said to be common to all human cultures; it can certainly be regarded as among the earliest

creations of any human society. This is because the material existence and survival of the human society depend on the ability of
man to make at least simple tools and equipment and to develop
techniques essential for the production of basic human needs such
as food, clothing, shelter, and security. The concern for such
needs was naturally more immediate than the pursuit and acquisition of the systematic knowledge of nature—that is, science. Thus,
in all human cultures and societies the creation of simple forms of
technology antedates science—the rational and systematic pursuit
of knowledge and understanding of the natural world, of the
processes of nature, based on observation and experiment. The
historical and functional priority of technology over science was
also a phenomenon even in the cultures of Western societies, historically the home of advanced and sophisticated technology.
From antiquity on, and through the Middle Ages into the modern
European world, innovative technology showed no traces of the
application of consciously scientific principles.[14] Science-based
technology was not developed until about the middle of the nineteenth century.[15] Thus, technology was for centuries based on
completely empirical knowledge.

The empirical character of African thought in general and of its
epistemology in particular was pointed up in the preceding section. The pursuit of empirical knowledge—knowledge based on
experience and observation, and generally oriented towards the
attainment of practical results—underpinned a great deal of the
intellectual enterprise of the traditional setting. (Note that philosophical knowledge was also thought to have a practical orientation.) And so, like other cultures of the world, practical knowledge
and the pursuit of sheer material well-being and survival led the
cultures of Africa to develop technologies and techniques, simple
in their forms, as would be the case in premodern times. Basic
craftsmanship emerged: farming implements such as the cutlass,
hoe and axe were made by the blacksmith; the goldsmith produced
the bracelet, necklace, and rings (including the ear-ring): 'African
coppersmiths have for centuries produced wire to make bracelets
and ornaments—archaeologists have found the draw-plates and
other wire-making tools.'[16] There were carpenters, wood-carvers,

[14] Lynn White, *Medieval Religion and Technology: Collected Essays*
(Berkeley: University of California Press, 1978), p. 127.
[15] Lord Todd, *Problems of the Technological Society*, The Aggrey-
Fraser-Guggisberg Memorial Lectures (Published for the University of
Ghana by the Ghana Publishing Corporation, Accra, 1973), p. 8.
[16] Arnold Pacey, *The Culture of Technology* (Oxford: Basil Blackwell,
1983), p. 145.

potters, and cloth-weavers, all of whom evolved techniques for achieving results. Food production, processing and preservation techniques were developed, and so were techniques for extracting medicinal potencies from plants, herbs and roots. A number of these technical activities in time burgeoned into industries.

There was, needless to say, a great respect and appreciation for technology because of what it could offer the people by way of its products. The need for, and the appreciation of, technology should have translated into real desire for innovation and improvement on existing technological products and techniques. There is, however, not much evidence to support the view that there were attempts to innovate technologies and refine techniques received from previous generations. There were no doubts whatsoever about the potencies of traditional medicines extracted from plants and herbs—the basis of the health care delivery system in the traditional setting and, to a very great extent, in much of rural Africa today. Yet, there were—and are—enormous problems about both the nature of the diagnosis and the appropriate or reliable dosage, problems which do not seem to have been grappled with. Diagnosis requires systematic analysis of cause and effect, an approach which would not be fully exploited in a system, like the one evolved by our cultures, which would often explain the causes of illness, as they would many other natural occurrences, in agentive (i.e. supernatural, mystical) terms, as I explained in the previous section. Such a causal approach to coping with disease would hardly dispose a people towards the search for effective diagnostic technologies.

Traditional healers were often not short on prescriptive capabilities: they were capable in a number of cases of prescribing therapies often found to be efficacious. But their methods here generated two problems: one was the preparation of the medicine to be administered to a patient; the other was the dosage—the quantity of the medicine for a specific illness. Having convinced himself of the appropriate therapeutic for a particular disease—a therapeutic which would often consist of a concoction, the next step for the herbal healer was to decide on the proportion (quantity) of each herbal ingredient for the concoction. Second, a decision had to be taken on the appropriate and effective dosage for a particular illness. Both steps obviously required exact measurement of quantity. The failure to provide exact measurement would affect the efficacy of the concoction as well as the therapeutic effect of the dosage; in the case of the latter, there was the possibility of under-dosage or over-dosage. Yet the need for exact measurement does not seem to have been valued and pursued by our cultures, a cultural defect which in fact is still taking its toll also in the maintenance of machines by our mechanics of today. Wiredu mentions the case of a Ghanaian mechanic who, in working on engine maintenance, would resort to the use of his sense of sight rather than of a feeler gauge in adjusting the contact breaker point in the distributor of a car.[17] The mechanic, by refusing to use a feeler gauge and such other technical aids, of course fails to achieve the required precision measurement. When it is realized that the habit or attitude of the mechanic was not peculiar to him but that it is a habit of a number of mechanics in our environments, it can be said that the development of that habit is a function of the culture. If one considers that precision measurement is basic not only for the proper maintenance of machines, but also for the quality of manufactured products of all kinds, one can appreciate the seriousness of the damage to the growth of technology caused by our cultures' failure to promote the value of precision measurement.

Even though it is true to say that historically technology was for centuries applied without resort to scientific principles, it is also conceivable that this fact must have slowed down the advancement of technology. It deprived technology of a necessary scientific base. The making of simple tools and equipment may not require or rest on the knowledge of scientific principles; but not so the pursuit of most other technological enterprises and methods. It cannot be doubted that the preparation of medicinal concoctions by traditional African herbal healers and their prescriptive dosages, for instance, must have been greatly hampered by the failure to attend to the appropriate scientific testing of the potencies of the various herbs and the amounts of each (herb or plant) required in a particular concoction. Theoretical knowledge should have been pursued to complement their practical knowledge.

Food technology, practised in the traditional setting mainly by African women, was a vibrant activity, even though the scientific aspect of it was not attended to. According to Sefa-Dede, who has done an enormous amount of research in traditional food technology in Ghana, 'The scientific principles behind the various unit operations may be the same as found in modern food technologies, but the mode of application may be different.'[18] The techniques traditionally deployed in food preservation undoubtedly involve the application of principles of science: physics, chemistry, and biology, which the users of those techniques may not have been

[17] Kwasi Wiredu, *Philosophy and an African Culture* (Cambridge University Press, 1980), p. 15.

[18] S. Sefa-Dede, 'Traditional Food Technology', in R. Macrae, R. Robinson and M. Sadler (eds), *Encyclopedia of Food Science, Food Technology and Nutrition*, (New York: Academy Press, 1993), p. 4600.

aware of. The techniques of preserving food all over Africa include drying, smoking, salting, and fermenting. The drying technique is aimed at killing bacteria and other decay-causing micro-organisms and thus preserving food intact for a long time; smoking serves as a chemical preservative; and so does salting which draws moisture and micro-organisms from foods; fermentation of food causes considerable reduction of acidity levels and so creates conditions that prevent microbial multiplication.[19] It is thus clear that there are scientific principles underlying these methods.

Let us take the case of a woman in the central region of Ghana underlying whose practice of food technology is clearly a knowledge of some principles of physics, chemistry and metallurgy. The woman in question is a processor of *fante kenkey*, a fermented cereal dumpling made from maize. Maize dough is fermented for two to four days. A portion of the dough is made into a slurry and cooked into a stiff paste. This is mixed with the remaining portion through a process called aflatization to produce aflata. This is wrapped in dried banana leaves and boiled for three to four hours until it is cooked. Now, this woman is able to solve a problem arising from the technique she uses in processing *fante kenkey*, to the amazement of a modern scientific research team interested in studying traditional food technology.[20] The woman challenged the research team to indicate how they could solve a very practical problem which can arise when one is boiling *fante kenkey* in a 44-gallon drum. This was the problem:

> Imagine that you have loaded a 44-gallon barrel with uncooked *fante kenkey*. You set the system up on the traditional cooking stove, which uses firewood. The fire is lit and the boiling process starts. In the middle of the boiling process, you notice that the barrel has developed a leak at its bottom. The boiling water is gushing into the fire and gradually putting off the fire. What will you do to save the situation?

The possible solutions suggested by the research team were found

[19] S. Sefa-Dede, ibid., also, 'Harnessing food technology for development', in S. Sefa-Dede and R. Orraca-Tetteh (eds) *Harnessing Traditional Food Technology for Development* (Department of Nutrition and Food Science, University of Ghana, Legon, 1989); Esi Colecraft, 'Traditional food preservation: an overview', *African Technology Forum*, Vol. 6, No. 1, (Feb./March, 1993), pp. 15–17.

[20] The encounter was between this traditional woman food technologist and research scientists and students from the Department of Nutrition and Food Science of the University of Ghana headed by Professor S. Sefa-Dede. The account of the encounter presented here was given to me by Sefa-Dede both orally and in writing, and I am greatly indebted to him.

to be impractical. One solution given by the team was to transfer the product from the leaking barrel into a new one. There are at least two reasons why this could not be done: one was that the *kenkey* will be very hot and difficult to unpack; the process will also be time-consuming; another reason was that another barrel may not be available.

The traditional woman food technologist then provided the solution: Adjust the firewood in the stove to allow increased burning; then collect two or three handfuls of dry palm kernel and throw them into the fire—this will heat up and turn red hot; finally, collect coarse table salt and throw it into the hot kernels. The result will be that the salt will explode and in the process seal the leak at the bottom of the barrel.

According to Sefa-Dede, the solution provided by the woman is based on the sublimation of the salt with the associated explosion. The explosion carries with it particles of salt which fill the opening. It is possible that there is interaction between the sodium chloride in the salt and the iron and other components forming the structure of the barrel. A few questions may arise as one attempts to understand the source of knowledge of the traditional practitioners: Why was dry palm kernel used as heat exchange medium? What is peculiar about table salt (sodium chloride) in this process? In the case under discussion, it can certainly be said that the woman has some knowledge about the thermal properties of palm kernels. (It is possible that there is traditional knowledge about the excellent heat properties of palm kernels. For, traditional metal smelters, blacksmiths, and goldsmiths are known to use palm kernels for heating and melting various metals). The woman, it can also be said, does have added knowledge of some chemistry and metallurgy. Even though it is clear that the ideas and solutions which the woman was able to come up with are rooted in basic and applied scientific principles, she cannot, like most other traditional technology practitioners, explain and articulate those principles. But not only that: they must have thought that the whys and hows did not matter: it was enough to have found practical ways to solve practical problems of human survival.

Thus, the pursuit of the principles would not have been of great concern to the users of traditional technologies, concerned, as they were, about reaping immediate practical results from their activities. The result was that there was no real understanding of the scientific processes involved in the technologies they found so useful. Yet, the concern for investigating and understanding those principles would most probably have led to innovation and improvement of the technologies. It can therefore be said that the weak scientific base of the traditional technology stunted its growth, and accounted for the maintenance and continual practice of the same old techniques. The understanding of the principles involved would probably have generated extensive innovative practices and the application of those principles to *other* yet-unknown technological possibilities. It clearly appears to be the case that once some technique or equipment was known to be working, there was no desire or enthusiasm on the part of its creators or users to innovate and improve on its quality, to make it work better or more efficiently, to build other—and more—efficient tools. Was this sort of complacency, or the feeling of having reached a cul-de-sac or of having come to the end of one's intellectual or technological tether, a reflection on the levels of capability that could be attained by our cultures?

Approaches to Developing A Modern Technology

It can hardly be denied that technology, along with science, has historically been among the central pillars—as well as the engines—of modernity. It is equally undeniable that the modern world is increasingly becoming a technological world: technology is, by all indications, going to become the distinguishing feature of global culture in the coming decades. Africa will have to participate significantly in the cultivation and promotion of this aspect of human culture, if it is to benefit from it fully. But the extensive and sustained understanding and acquisition of modern technology insistently require adequate cultivation of science and the scientific outlook. The acquisition of scientific and technological outlook will in turn require a new mental orientation on the part of the African people, a new and sustained interest in science to provide a firm base for technology, a new intellectual attitude to the external world uncluttered by superstition, mysticism and other forms of irrationality; the alleged spirituality of the African world, which was allowed in many ways to impede sustained inquiries into the world of nature, will have to come to terms with the physical world of science. Knowledge of medicinal plants, for instance, qua scientific knowledge, must be rescued from the quagmire of mysticism and brought in to the glare of publicity, and its language made exoteric and accessible to many others.

The need for sustained interest in science is important for at least two reasons: to provide an enduring base for a real technological take-off at a time in the history of the world when the dynamic connections between science and technology have increasingly been recognized and made the basis of equal attention to both: technology has become science-based, while science has become technology-directed. The second reason, a corollary of the first, is that it is the application of science to technology that will help improve traditional technologies.

Ideally, technology, as a cultural product, should take its rise from the culture of a people, if it is to be directly accessible to a large section of the population and its nuances fully appreciated by them. For this reason, one approach to creating modern technology in a developing country is to upgrade or improve existing traditional technologies whose developments, as I have already indicated, seem to have been stunted in the traditional setting because of their very weak science base. Let us recall the case of the woman food technologist referred to in the previous section. She was able to find practical ways of solving problems that emerged in the course of utilizing some technology by resorting to ideas and solutions which are obviously rooted in basic science, but without the benefit of the knowledge of chemistry, physics, engineering, or metallurgy. From the technology she used, questions that arise would include the following: Why were dry palm kernels used as heat exchange medium? What is peculiar about table salt (sodium chloride) in this process? Yet for most traditional technology practitioners the whys and how did not often matter, so long as some concrete results can be achieved through the use of a particular existing technology. But the why and how questions of course do matter a great deal. Improving traditional technologies will require not only looking for answers to such questions, but also searching for areas or activities to which the application of existing technologies (having been improved) can be extended.

Traditional technologies have certain characteristics which could—and must—be featured in the approach to modern technology in a developing country. Traditional technologies are usually simple, not highly specialized technologies: this fact makes for the involvement of large numbers of the people in the application or use of the technologies, as well as in their development; but it also promotes indigenous technological awareness. The materials that are used are locally available and the processes are effective. (In the case of the woman's food technology the materials in question, namely, palm kernels and table salt, are household items which are readily available.) Traditional technologies are developed to meet material or economic needs—to deal with specific problems of material survival. They can thus be immediately seen both as having direct connections with societal problems and as appropriate to meeting certain basic or specific needs. If the technologies that will be created by a developing country in the modern world feature some of the characteristics of the traditional technologies, they will have greater relevance and impact on the social and economic life of the people.

The improvement of traditional technologies is contingent on at least two factors. One is the existence or availability of autonomous, indigenous technological capacities. These capacities would need to be considerably developed. The development of capacities in this connection is not simply a matter of acquiring skills or techniques but, perhaps more importantly, of understanding, and being able to apply, the relevant scientific principles. It might be assumed that the ability to acquire skills presupposes the appreciation of scientific principles; such an assumption, however, would be false. One could acquire skills without understanding the relevant underpinning scientific principles. The situation of the woman food technologist is a clear case in point. However, the lack of understanding of the relevant scientific principles will impede the improvement exercise itself. The other factor relates to the need for change in certain cultural habits and attitudes on the part of artisans, technicians, and other practitioners of traditional technologies. Practitioners of traditional technologies will have to be weaned away from certain traditional attitudes and be prepared to learn and apply new or improved techniques and practices. Some old, traditional habits, such as the habit, referred to in the previous section, of the automatic use of the senses in matters of precision measurements, will have to be abandoned; adaptation to new—and generally more effective—ways of practising technology, such as resorting to technical aids in precision measurements, will need to be pursued. It is the cultivation of appropriate attitudes to improved—or modern—technology and the development of indigenous technological capacity that will provide the suitable cultural and intellectual receptacle for the modern technologies that may be transferred from the technologically advanced industrial countries of the world to a developing country.

Now, the transfer of technology from the technologically developed world is a vital approach to bringing sophisticated technology to a developing country. It could also be an important basis for developing, in time, an indigenous technological capacity and the generation of fairly advanced indigenous technologies. But all this will depend on how the whole complex matter of technology transfer is tackled. If the idea is not executed well enough—if it is bungled—it may lead to complacency and passivity on the part of the recipients, reduce them to permanent technological dependency, and involve them in technological pursuits that may not be immediately appropriate to their objectives of social and economic development. On the other hand, an adroit approach to technology transfer by its recipients will, as I said, be a sure basis for a real technological take-off for a developing country.

Transfer of technology involves taking some techniques and practices developed in some technologically advanced country to some developing country. The assumption or anticipation is that the local people, i.e. the technicians or technologists in the developing country, will be able to acquire the techniques transferred to them. Acquiring techniques theoretically means being able to learn, understand, analyse, and explain the whys and hows of those techniques and thus, finally, being able to replicate and design them off the local technologist's own bat. It is also anticipated that the local technologist, who is the beneficiary of the transferred technology, will be able not only to adapt the received technology to suit the needs and circumstances of the developing country, but also to build on it and, if the creative capacity is available, to use it as an inspiration to create new technologies appropriate to the development requirements and objectives of the developing country.

The assumptions and anticipations underlying the transfer of technology of course presuppose the existence, locally, of an autonomous technological capacity which can competently deal with, i.e. disentangle, the intricacies of the transferred technology. In the event of the nonexistence of an adequate indigenous technological capacity, the intentions in transferring technology will hardly be achieved. There is some kind of paradox here: autonomous, indigenous technological capacities are expected to be developed *through* dealing with transferred technologies (this is certainly the ultimate goal of technology transfer); yet, the ability to deal effectively with transferred technologies requires or presupposes the existence of indigenous technological capacities adequate for the purpose. However, the paradox can be resolved if we assume that the indigenous technological capacities will exist, albeit of a minimum kind, which would, therefore, need to be nurtured, developed, and augmented to some level of sophistication required in operating a modern technology. The assumptions also presuppose that the transferred technology, developed in a specific cultural milieu different in a number of ways from that of a developing country, is easily adaptable to the social and cultural environment of the developing country. This presupposition may not be wholly true. However, despite the problems that may be said to be attendant to the transfer of technology, technology transfer is, as I said, an important medium for generating a more efficient modern technology in a developing country.

Now, technology is of course developed within a culture; it is thus an aspect—a product—of culture. Technology transfer, then, is certainly an aspect of the whole phenomenon of cultural borrowing or appropriation which follows on the encounters between cultures. There appears, however, to be a difference between transfer of technology to a developing country and the normal appropriation by a culture of an alien cultural product. The difference arises because of the way the notion of technology transfer is conceived and executed. It can be admitted that what is anticipated in technology transfer is primarily *knowledge* of techniques, methods and materials all of which are relevant to matters of industrial production. But knowledge is acquired through the active participation of the recipient; it is not transferred on to a passive agent or receptacle. In the absence of adequate and extensive knowledge and understanding of the relevant scientific principles, the attitudes of the recipients of transferred technology will only be passive, not responsive in any significant way to the niceties of the new cultural products being introduced to them. In the circumstance, that which is transferred will most probably remain a thin veneer, hardly affecting their scientific or technological outlook and orientation. Machines and equipment can be transferred to passive recipients who may be able to use them for a while; but the acquisition of knowledge (or, understanding) of techniques—which is surely involved in the proper meaning of technology—has to be prosecuted *actively*, that is, through the active exercise of the intellects of the recipients.

In an ideal situation of cultural borrowing, an element or product of the cultural tradition of one people is accepted and taken possession of by another people. The alien cultural product is not simply 'transferred' to the recipients. Rather, goaded by their own appreciation of the significance of the product, they would seek it, acquire it, and appropriate it, i.e. make it their own; this means that they would participate actively and purposefully in the acquisition of the product. To the extent, (i) that what is called technology transfer is an aspect of the phenomenon of cultural borrowing, and (ii) that the people to whom some technology is transferred are, thus, expected to understand and take possession of it through active and purposeful participation in its acquisiton, 'transfer of technology' is, in my view, a misnomer. For, what is transferred may not be acquired, appropriated, or assimilated.

For the same reasons, Ali Mazrui's biological metaphor of 'technology transplant' will not do either. In Mazrui's view, 'there has been a considerable amount of technology transfer to the Third World in the last thirty years—but very little technology transplant. Especially in Africa very little of what has been transferred has in fact been successfully transplanted.'[21] To the extent, (i) that this biological or medical metaphor clearly involves passivism on the part of the recipient (i.e. the patient), who thus has no choice in actively deciding the 'quality' of the foreign body tissue to be sewn onto his own body, and (ii) that there is no knowing whether the physical constitution of the recipient will accept or reject the new body tissue, the biological perception of acquiring the technological products of other cultures is very misleading. On a further ground, the biological metaphor will not do: the body onto which a foreign body tissue is to be transplanted is in a diseased condition which makes it impossible for it to react in a wholly positive manner to its new 'addition' and to take advantage of it; even if we assume, analogically, that the society that is badly in need of the technological products of other cultures is technologically or epistemically 'diseased', the fact would still remain that, in the case of the human society, the members of the society would, guided by their needs, be in a position not only to decide on which technological products of foreign origin they would want to acquire, but also to participate actively and positively in the appropriation of those products.

Thus, neither technology transfer nor technology transplant is a fruitful way of perceiving—and pursuing—the acquisition of technology from other cultures; neither has been a real feature or method in the phenomenon of cultural borrowing. Our historical knowledge of how the results of cultural encounters occur seems to suggest the conviction that what is needed is, not the transfer or transplant of technology, but the *appropriation* of technology—a perception or method which features the active, adroit, and purposeful initiative and participation of the recipients in the pursuit and acquisition of a technology of foreign production.

It must also be noted that just as in cultural borrowing there are surely some principles or criteria that guide the borrowers in their selection of products from the alien—i.e. the encountered—culture, so, in the appropriation of technology some principles or criteria would need to be established to guide the choice of the products of technology created in one cultural environment for use in a different environment. Technology can transform human society in numerous ways. For this reason, a developing country will have to consider technology rather as an instrument for the realization

[21] Ali A. Mazrui, 'Africa between ideology and technology: two frustrated forces of change', in Gwendolen M. Carter and Patrick O'Meara (eds), *African Independence: The First Twenty-Five Years* (Bloomington: Indiana University Press, 1985), pp. 281–282.

of *basic* human needs than as an end—as merely a way of demonstrating human power or ingenuity. The word 'basic' is important here and is used advisedly: to point up the need for technology to be concerned fundamentally and essentially with such human needs as food, shelter, clothing, and health. The pursuit and satisfaction of these basic needs should guide the choice and appropriation of technology. Thus, what ought to be chosen is the technology that will be applied to industry, food and agriculture, water, health, housing, road and transportation, and other most relevant activities that make ordinary life bearable. On this showing, military and space exploration technologies, for instance, may not be needed by a developing country. However, as a developing country comes to be increasingly shaped by technology, certain aspects of technology will become a specialized knowledge; it will *then* become necessary to create a leaven of experts to deal with the highly specialized aspects of those technologies.

The adaptability of technological products to local circumstances and objectives must be an important criterion in the appropriation and development of technology.

Finally, the fundamental, most cherished values of a culture will also constitute a criterion in the choice of technology. Technology, I said, can transform human society. This social transformation will involve changes not only in our ways and patterns of living, but also in our values. But human beings will have to decide whether the (new) values spewed out by technology are the kinds of values we need and would want to cherish. Technology emerges in, and is fashioned by, a culture; thus, right from the outset, technology is driven or directed by human purposes, values and goals. And, if this historical relation between technology and values is maintained, what will be produced for us by technology will (have to) be in consonance with those purposes, values and goals. Technology was made by man, and not man for technology. This means that human beings should be the centre of the focus of the technological enterprise. Technology and humanism (i.e. concern for human welfare) are—and should not be—antithetical concepts; technology and industrialism should be able to coexist with the concern for the interests and welfare of the people in the technological society. So, it should be possible for an African people to embark on the 'technologicalization' of their society without losing the humanist essence of their culture. The value of concern for human well-being is a fundamental, intrinsic and self-justifying value which should be cordoned off against any technological subversion of it. In this connection, let me refer to the views expressed by Kenneth Kaunda in the following quotation:

I am deeply concerned that this high valuation of Man and respect for human dignity which is a legacy of our (African) tradition should not be lost in the new Africa. However 'modern' and 'advanced' in a Western sense the new nations of Africa may become, we are fiercely determined that this humanism will not be obscured. African society has always been Man-centred. We intend that it will remain so.[22]

I support the view that the humanist essence of African culture—an essence that is basically moral[23]—ought to be maintained and cherished in the attempt to create modernity in Africa. It must be realized that technology alone cannot solve all the deep-rooted social problems such as poverty, exploitation, economic inequalities, and oppression in human societies *unless* it is underpinned and guided by some basic moral values; in the absence of the strict application of those values, technology can in fact create other problems, including environmental problems. Social transformation, which is an outstanding goal of the comprehensive use of technology, cannot be achieved unless technology moves along under the aegis of basic human values. Technology is a human value, of course. And because it is basic to the fulfillment of the material welfare of human beings, there is a tendency to privilege it over other human values. But to do so would be a mistake. The reason is that technology is obviously an instrument value, not an intrinsic value to be pursued for its own sake. As an instrument in the whole quest for human fulfillment, its use ought to be guided by other—perhaps intrinsic and ultimate—human values, in order to realize its maximum relevance to humanity.

In considering technology's aim of fulfilling the material needs of humans, the pursuit of the humanist and social ethic of the traditional African society can be of considerable relevance because of the impact this ethic can have on the distributive patterns in respect of the economic goods that will result from the application of technology: in this way, extensive and genuine social—and in the sequel, political—transformation of the African society can be ensured, and the maximum impact of technology on society achieved.

[22] Kenneth Kaunda, *A Humanist in Africa* (New York, 1966), p. 28.
[23] Gyekye, *An Essay on African Philosophical Thought*, pp. 143–146.

Art and Technology: An Old Tension

ANTHONY O'HEAR

This is not the first time the title 'Art and Technology' has been used, but to distinguish what I have to say from Walter Gropius's Bauhaus exhibition of 1923, I am subtitling my paper 'an old tension', where the architect spoke of 'a new unity'. In a way, Gropius has been proved right; the structures of the future avoiding all romantic embellishment and whimsy, the cathedrals of socialism, the corporate planning of comprehensive Utopian designs have all gone up and some come down. We have a mass media culture also largely made possible by technology. Corporatist architecture, whether statist 'social housing' or freemarket inspired, films, videos, modern recording and musical techniques are all due to technological advances made mostly this century. Only in a very puritanical sense could what has happened be thought of as inevitably bringing with it enslavement. All kinds of possibilities are now open to artists and architects, which would have been imaginable a few decades ago. No one is forced to use these possibilities in any specific way.

If I have a complaint about what has happened in the arts this century, it is not that technology is constraining imagination, or making slaves of us. It is rather that it isn't constraining it enough, that technology is removing those very constraints which made art a matter of craft, rather than an unfettered display of expression and imagination.

It was common at one time to praise the avant-garde architects of the first half of the century for (in Pevsner's words) 'courageously' breaking with the past, and accepting 'the machine age in all its implication: new materials, new processes, new forms, new problems'.[1] We are now in a better position than Pevsner was in 1943 to judge whether this courage was advisable or not: whether it was something forced on the artist or architect of integrity by new materials, etc., or whether it was, rather, a question of the possibilities offered by new materials leading to the deformation of an old craft. Certainly this century we have had an aesthetic of functionality which intentionally broke with the past and in so doing produced buildings which were by any normal human standard uncomfortable and inconvenient, precisely because they

[1] N. Pevsner, *An Outline of European Architecture*, 5th edition, (Harmondsworth: Penguin, 1957), p. 281.

swept away much ornamentation and detail whose role we now know was not merely decorative.

It is part of modernist propaganda that the aesthetic of the Bauhaus and Le Corbusier was forced on us by new materials and techniques. That this is false is shown by the fact that there are architects today who are using those very materials and techniques to produce a new classicism, just as Giles Gilbert Scott earlier this century had used reinforced concrete to largely unmodernist effect. In particular, Scott refused to expose concrete, regarding it as visually crude.[2] Modernists would doubtless accuse Scott of being untrue to his materials, but with decades of what David Watkin has called the 'graceless weathering' of concrete, that doctrine is itself looking a little tawdry to-day.[3] It is, in any case, an *aesthetic* doctrine, not one forced on us by materials or by technology. We are perfectly free to reject its sillier and uglier implications on aesthetic grounds, as indeed Ruskin, its original proponent, would have done. Ruskin was seeking not to do away with all ornamentation, but only with what he regarded as the dishonesty of certain types of ornamentation, all of which goes to show once one gets away from the simplicities of modernistic sloganizing, even the meaning of the doctrine of truth to materials is far from clear.

The artist, according to Ruskin,

> is pre-eminently a person who sees with his eyes, hears with his ears, and labours with his body, as God constructed them; and who, in using instruments, limits himself to those which convey or communicate his human power, while he rejects all that increase it. Titian would refuse to quicken his touch by electricity; and Michael Angelo to substitute a steam-hammer for his mallet. Such men not only do not desire, they imperatively and scornfully refuse, either the force, or the information, which are beyond the scope of the flesh and the senses of humanity.[4]

[2] See David Watkin, *A History of Western Architecture* (London 1986), p. 557.

[3] Ibid. p. 558.

[4] John Ruskin, *Deucalion* (1875–83), Vol. I, Ch. II, Section 4.

In writing this paper. I found myself increasingly drawn to Ruskin, and to what he has to say on the subjects of art and technology. This paper is not a paper on Ruskin, nor I do not accept everything Ruskin said on everything. (How could I when Ruskin is *notoriously* inconsistent?) But even though Ruskin is not normally regarded as a philosopher, what I write could be nevertheless regarded as a philosophical homage to our greatest writer on art (not that he would have been pleased at being so described).

Unlike Ruskin, I am not confident about what Titian or Michelangelo might or might not have done, had they lived the age of electricity, or of the steam-hammer, or of acrylic paint, or of computer-assisted design, or of epoxy resin. But what I do know is that with all our technological advances, there is no one painting today who can convey colour, shade, flesh or cloth as Titian and dozens of his contemporaries could. It is also significant in this context how our supposedly advanced 'scientific' restoring techniques have in the case of Titian often simply had the effect of removing the glazes and varnishes on which his flesh tones, shading, perspectives and half-tints depended: compare, for example, the ruin of 'Venus, Cupid and an Organist' in the Prado, or the garish 'St Margaret' from the Heinz Kisters collection with the uncleaned 'Annunciation' from Naples.[5] And it is hard to think of any sculptor of note today even interested in carving forms of classical beauty from blocks of marble, let alone worth mentioning alongside Michelangelo.

If Titian and Michelangelo belonged to a technologically primitive age, it is as if their technological limitations simply spurred on their mastery of technique, whereas our technological advancement has led to a loss of technique. Technique is certainly central to art, for reasons we will come to. Indeed, one of the unfortunate effects of the entry of technology into the world of art has been the downgrading of technique and the upgrading of the idea. The thought is: why bother with painting when I can get what I want much more easily and quickly with a camera? Why should I run the fearsome risks inherent in carving, in which a few bad strokes can ruin the integrity of my stone, when I can do what Michelangelo did not have available, that is model in clay and infallibly point-up my block of stone (or even get an assistant to point-up for me)? Why should architects learn to draw the human figure—knowledge which for centuries formed the basis of a feeling for architectural proportion and sensibility—when computer-assisted design can apparently make their lives so much more exciting?

Computer-generated graphics and designs now take us far beyond the static camera or the nineteenth-century sculptor's pointing machine, allowing the instant realization of complex and many-layered conceptions, human input into the result being infinitesimal compared to what is required in painting even the

[5] All this was revealed in the great exhibition of Titian and his contemporaries in the Grand Palais in Paris in 1993. See the catalogue of that exhibition, *Le Siècle de Titien* (Paris, 1993), especially illustrations 176, 250, 251.

most formulaic canvas. From what appears to be the opposite end of the artistic spectrum, Richard Long—who is paid large sums of money to walk from one place to another and take photographs as he goes—says that he likes 'the simplicity of walking, the simplicity of stones ... common means given the simple twist of art ... *sensibility without technique*'.[6] This sort of conceptual art may seem a far cry from the high-tech of the studios of contemporary architects and video producers, but in both cases we have lost the ancient and noble idea of art as technē or craft, in which all that was done was done by human hand and controlled by human intelligence.

Plato, in common with his fellow Athenians of the fifth century BC, put painters in the same category as shipwrights, builders and 'other craftsmen', that is people who mould their materials, with craft or artistry (or technē), fitting each bit to harmonize with the others 'until they have combined the whole into something well ordered and regulated'.[7] Skill (technē) is needed for art, but over and above that, some imaginative power to order and co-ordinate the parts in a manner appropriate to the intended effect. Part of Plato's point in discussing painting is to contrast activities, which make use of a recognizable technē, with poetry, which according to Plato does not, relying instead on an uncertain and morally unreliable inspiration (and is indeed the sensibility without technique admired by Long). Nevertheless, Plato's prejudice against poets aside, the linking of art with crafts such as building is a healthy antidote against that form of idealism regarding art which technology encourages. We need to remind ourselves of the way in which art grows out of our embodiment and affects us via our senses.

Where we come to the work of a true artist, we get a full-blooded humanism:

> The day's work of a man like Mantegna or Paul Veronese consists of an unfaltering, uninterrupted series of movements of the hand more precise than those of the finest fencer: the pencil leaving one point and arriving at another, not only with unerring precision at the extremity of the line, but with an unerring and yet varied course—sometimes over spaces a foot or more in extent—yet a course so determined everywhere, that either of these men could, and Veronese often does, draw a finished profile, or any other portion of the face, with one line, not afterwards changed. Try, first, to realise to yourselves the muscular precision of the action, and the intellectual strain of it: for the

[6] Richard Long, *Words after the Fact*, text in his 1980 catalogue for the d'Offay gallery, London.

[7] Plato, *Gorgias*, 503e.

movement of a fencer is perfect in practiced monotony; but the movement of the hand of a great painter is at every instant governed by a direct and new intention.[8]

It is because *everything* is a craft-based work of art is governed by a direct and new intention, that a painting is potentially of far more critical and aesthetic interest than a photograph say. In a photograph, the photographer governs and intends only the general shape and outline depicted, and the outcome is not the result of a constant and continuous reflective interaction with the material over a long period.

In creating a work of art the artist occupies a dual role. He is both maker and audience, or, rather as maker, he constantly puts himself in the position of his audience, judging for himself just how what he makes will appear to the audience. In the light of this knowledge, he will modify what he does (and sometimes, as in the case of Bruckner, he will modify what he does in the light of actual audience response).

If the creator of a work of art puts himself in the position of his imagined audience, the audience attends to a work of art not as to a natural phenomenon (such as a cloud, a crystal, or an untamed landscape), but as to something driven by human intelligence and expressive of human emotion. In so doing, the audience is confident that it is not indulging in the exercise of pathetic fallacy, that is, it is not imputing to unplanned and unintended phenomena characteristics which are properly attributed only to planned and intended things. A work of art is produced by a human being in order to express some vision and set of intentions its creator has in making it. As such, a work of art is of interest for what it reveals about the human world, the world in which intentions, institutions and traditions introduce meaning into an otherwise meaningfully empty universe. As located in the human world, works of art are liable to the same types of evaluation as other human works and deeds: 'there is no moral vice, no moral virtue which does not have its precise prototype in the art of painting', says Ruskin,[9] and the same can be said of the human attributes other than the moral. (It does not, incidentally follow from this that the *artist* who gives expression to, say, modesty or chastity is himself chaste or modest: that there is always (?) an element of theatricality about art does not mean that a Raphael or a Wagner might not understand human feelings very far from their own behaviour and evoke them in their work.)

[8] John Ruskin, *Lectures on Art* (1870), Lecture III, Section 71.
[9] John Ruskin, *Elements of Drawing* (1857), Section 135.

What, though, would we say were we to do a Turing test on a work of art, and discover that something we had imagined to be the work of human artist and assessed as such, had in fact been generated by a computer? This question takes us to the core of the issue which concerns us, for it raises in a dramatic way the connection between the work of art as a product, and its history or mode of production. What I am going to argue for is for a form of externalism regarding works of art: that their visible form or intelligible content is not all they are, that part of what makes them what they are is in fact that they have been produced in a particular way, as a result of human intentional activity, one aim of which is to produce a realization in the audience that the artist has intended that the audience have an experience of a particular sort.

This does not mean that there cannot be failures in an artistic attempt to express a particular vision, nor that the result may not be ambiguous in various ways, and so fall below or above the sense the artist has as to what he is doing. But that we can judge a work to have failed against what we take to be the artist's intention is a justifiable and often illuminating commonplace of art criticism, but one which is to the point only where a human action is concerned. Similarly, that a work of art is ambiguous in various ways may tell us something about hidden aspects of ideas or feelings—hidden even to the original proponent of them—but which emerge precisely in executing the intention to articulate the surface feelings artistically.

In exploring the relation between works of art and their producers, we need initially to distinguish between production and reproduction. There are certainly works of art which can be mechanically replicated, even after their creators have lost interest in them, or even died. Sculptural bronzes, engravings and prints can all be produced in the absence of the originality artist, according to the matrix he has produced. (I leave aside the practice at least some sculptors and print-makers have of embellishing and refining the final results.) What, though, we have in these cases is mechanical reproduction or replication, rather than mechanical production. The artist is held to be responsible for the product, because he is taken to know how the final results will turn out, and to have designed his plate or model accordingly. We may speak here of mechanical replication rather than mechanical production, because in a sculptural bronze or an engraving the detail of every finished line and contour is or should be under the control of the artist: contrast that is with a photograph in which much of the detail inevitably has little to do with the photographer's intention, or even, in typical cases, his knowledge. (This, of course, contributes to the sense of *actualité* conveyed by many photographs, but it is *actualité* at the expense of art).

Reproduction of a less elevated sort can also be envisaged in the visual arts and in architecture, where an original work is simply copied. One could certainly imagine fairly expert copying of wholes or parts of originals by means of machines, without human intermediaries. Such replication might even survive the Turing test; that is observers might be unable to tell that a particular drawing or building was the computer-generated copy rather than the original. If computers were able to generate copies of existing works of art, it might well be possible for them to produce formulaic works which drew on repertoires derived from existing works, but which were not copies of any complete works. What, if anything, would follow for our concept of a work of art from this possibility?

What would follow would be this. Computer-generated works of the type indicated might be indistinguishable from human productions, but the meanings which were attributed to them would be due to their sharing in the forms of typical human productions. They would be parasitically meaningful, deriving their meanings from the techniques and conventions which human artists had developed in their works.

However, the possibility that a computer might produce an original work indistinguishable, at least in respect of its appearance, from something produced by a human being, does not show that we are wrong to see works of art as, in their primary sense, productions of human intention and intelligence, and to accord them the value we do accord them at least in part because of that. In particular, works of art, like other productions of the human world, derive their significance from their ability to express and incarnate the way things seem to members of human communities. They are also intended to evoke aesthetic, moral, intellectual or emotional reactions from audiences. As things do not *seem* at all to computers or other machines, and as computers and other machines do not react with feeling and sensitivity, or even with unfeeling or insensitivity, computer-produced art is viable only to the extent that computers are programmed to explicit knowledge and mimic and anticipate feelings which arise only within human experience.

Computer-generated art will in any case be unable to profit from that close and continuous monitoring of results in the light of human response which we spoke of earlier in connection with the glazing, shading and varnishing of the Venetian masters. It is hard to imagine that this knowledge could be algorithmically formalized and computed at all; works of art are, as Kant taught, essentially singular and individual. Their effects are not generalizable into other contexts, but dependent on the ungeneralizable judgement of both creator and perceiver.

We are now in a position to suggest two ways in which art cannot be mechanical:

(1) *The 'externalist' thesis, i.e. it matters whence artistic works derive*

Art is based on human life, its emotions, attitudes, feelings and institutions. While a purely machine-produced art, such as the fractal images produced by computers, may exemplify certain formally satisfying features, it necessarily lacks those references, explicit and implicit, to human life and sensibility, which make art more than a purely formal exercise. Art in the full sense is based in human experience. It does therefore matter where a work is grounded, and works not grounded directly in real human experience gain the honorific title of art only by being modelled on works which are. From such 'modelled' works we will not expect the same level or type of insight into the human condition as those that are created by human beings who have had the experiences relevant to their meaning, though we may learn something about the intentions and mentality of people in the art world who chose to exhibit them as though they were no different in interest from humanly produced works. It might also be a comment on a form of art, such as geometrical abstraction, which is *easily* mimiced by computers operating with little or no human guidance.

(2) *The singularity thesis*

Art of distinction is necessarily individual and unique. It impresses partly because in it we see the free exercise of skill and judgement so as to produce new effects and original works. The skill and judgement is that of the artist who moulds what he does, exercising his skill in the light of his judgement. It is a cliché, but nonetheless true, that there is barely a movement in any of Mozart's mature repertoire of symphonies, concerti and string quartets which is not full of surprises; truly a matter of, in Kant's terms, genius, the innate mental disposition through which, as Kant says, nature gives the rule to art.[10]. So a Mozartian computer could never write a Mozart work: like any pastiche, it would lack just those qualities of surprise, subtle melodic inventiveness, poignancy and control which characterize echt-Mozart.

[10] I. Kant, *Critique of Judgement* (1790), Section 46.

150

The singularity theses impinges on technology both as regards the need to experience a particular object and as regards ungeneralizable aspects of that object. In the first place as Kant insists, aesthetic judgments are judgments about single and particular objects. To make such a judgment, you have to experience the object in question. I can judge that, say, 'Bacchus and Ariadne' is a beautiful painting only if I am perceiving *it* or have perceived *it*. An account of the painting without the experience will not be sufficient. Even if I were able, dubiously, to list all its properties there would be no guarantee that in some other context, even one just slightly different from that produced by Titian, those very same properties might not produce a quite different and even an unbeautiful result. I have chosen my example with care. There are those—me included—who will hold that it is actually impossible to make a proper judgment about 'Bacchus and Ariadne' in 1995, because since 1962 when it was 'restored', Titian's painting no longer existed. The result is that 'in the view of many informed critics its systematic and unimaginative cleaning has resulted in a tragic disruption of the fragile total unity of the painting',[11] a view many will find confirmed if they go to the National Gallery in London today, and are struck both by the garish colours and the hard contours of the restoration. The influence of technology is double here: not only is it technology which has destroyed the soft lines and half tints of the original, but the harsh flattening aesthetic at work in the restoration is very much of that of the coloured photograph, the magazine and advertising.

What I say about restoration may seem to justify authentic performances of pre-nineteenth-century music, and to condemn any use of modern instruments in, say, Bach or Mozart. Actually, what I am saying has the opposite conclusion. It would clearly be wrong to play Bach as if he were a Liszt or a Brahms conjuring up waves of undifferentiated sound, or to fill in the sonorities of keyboard to Mozart's works without inserting the unwritten grace notes Mozart intends. But it does not follow that they cannot or should not be played on modern pianos. It is certainly arguable that Bach or Mozart were not exclusively aiming at the dead and staccato sounds characteristically associated with ancient keyboard instruments, and that the sonority and legato effects they clearly wanted can, by a Gould or a Perraiha, be better conveyed on a modern piano, without sacrificing clarity or lightness, than by a slavish search for authenticity. Our current search for musical authenticity

[11] Sarah Walden, *The Ravished Image* (London 1985), p. 139.
In what I have said about cleaning I have relied closely on Mrs Walden's excellent analysis.

could well be the contemporary equivalent of the National
Gallery's cleaning policies of the 1960s. The point is that in questions of art, what is required is judgment and insight respecting the
individual case, and that blanket decisions covering whole classes of
case (the technological approach) are likely to be wrong.

If aesthetic judgments are characteristically singular and ungeneralizable, the artist in producing a work will also be guided by
his 'genius', in Kant's terms, by a sense or talent for which no definite rule can be given.[12] What the artist must have is insight into
how something will work, what experiences it will produce in the
individual case. According to Ruskin, in the case of Veronese's
'Presentation of the Queen of Sheba' (Galleria Sabauda, Turin),
what might otherwise and in the other hands be a mess of 'trivial
and even ludicrous detail' in no way detracts from the nobleness of
the whole.[13] One could, of course, lay down a rule 'Detail is permitted (encouraged ?) to the extent that it doesn't detract, etc.,
etc.', but can there be a rule defining the appropriate extent? As
Kant again, has argued, a genius (a Veronese) can produce models,
which can serve as a standard or rule of judgment for others, but
doing this will not indicate 'scientifically' how to enable others to
produce similar products, if only because in matters of artistic
judgment and creation, the artist himself does not know how his
judgment works. What is at issue is the free play of the imagination, in which hitherto unassociated elements are brought together,
so as to open out the prospect of an 'illimitable field' of association
and representation. What is at issue is not a question of a closed
logical deduction nor of the systematic steps a scientist might take
in pursuing and testing a hypothesis.

A large part of what interests us in a work of art is an encounter,
not with a machine or an algorithm but with another human mind,
at once expressing itself freely, but also in control of what it is
doing. I am not saying that there might not be aesthetic interest in
a computer productions, in fractals and the like. There is aesthetic
interest in crystals, after all, and in landscape and cloud forma-
tions, an interest which does not diminish for the man unable to
see such things as God's handiwork.

But there is a different, and from the human point of view a
superior interest in recognizing an aesthetic phenomenon as produced by a human being. 'Its true delightfulness depends on our
discovering in it the record of thought and interest, and trials, and
heart breakings—of recoveries and joyfulness of success; all this
can be traced by a practised eye: but, granting it even obscure, it is

[12] See Kant, *Critique of Judgement*, Section 46.
[13] John Ruskin, *Modern Painters* (1860), Vol. 5, Part IX, Section 2.

presumed or understood; and in that is the worth of the thing, just as much as the worth of anything else we call precious.'[14] An artist works with the intention that an audience should, in its experience of the work, come to have just that experience. In coming to form the work in the light of the experience it is intended to evoke—a process which will itself refine both intention and experience as the artist works will, as Richard Wollheim has argued, be drawing on 'thoughts, beliefs, memories, and in particular emotions and feelings, that [he] had, and that, specifically caused him to create as he did'.[15] He will also, doubtless, be guided by notions of what he thinks appropriate, aesthetically, morally, commercially, spiritually and in a host of other ways. A work of art, then, as it is being produced draws on much of the fabric of its creator's life, outer and inner, private and public, and, at the same time, contributes to the development of that fabric.

If an artist is successful in his work, we can learn much about his inner life from contemplating is work, even without knowing much about his outer life. To put this another way, a work of art, if successful, draws on the artist's inner imaginative life and articulates aspect of it, and we learn something of that inner life from the works which stem from it, even without knowing much of the outer life in question. And, if Proust is to be believed, and the true life of an artist is that inner imaginative source on which he draws in his work, there can actually be disadvantages in knowing too much of the outer life, however much the outer life may have provided the bits and pieces which the inner self transforms. What is being said here is not that a work of art is or should be just a mirror of the feelings or biography of the artist. A good work of art involves a discipline—a technē—which will involve the artist in disassociating himself from his actual and current feelings. His work should, as T. S. Eliot insisted, be 'not a turning loose of emotion, but an escape from emotion ... not the expressions of personality, but an escape from personality'.[16] But, as Eliot adds, this impersonality is not something ungrounded in personal experience and passion; 'only those who have personality and emotions know what it means to escape from these things'. It is a consequence of this view that while art should not be read as pure autobiography, and while artists need not (and should not) aim at displaying the detail of their own personality, art which cannot in any

[14] John Ruskin, *Seven Lamps of Architecture* (1849), Ch. II, 'The Lamp of Truth', Section 19.

[15] R. Wollheim, *Painting as an Art* (London, 1987), p. 86.

[16] T. S. Eliot, 'Tradition and the Individual Talent', in *The Sacred Wood* (Methuen, 1960 edition), pp. 47–59, p. 58.

sense be regarded as successfully expressive of its creator's inner life and vision will, to that extent, be of less interest than that which can. Shakespeare and Homer are often regarded as the most impersonal of artists in that their work seems, by being so rich and deep and all-encompassing, to transcend their empirical existences and personalities. One could say, on the contrary, that part of what we admire about them is the manifestation of an inner life and vision which was so comprehensive and full of insight.

I believe that it would be nonsense to think of a computer having an inner life in this sense, or of its inner life being composed through its reactions to an outer world, including the human world, but even if a computer did have an inner life, it would not be a human or even an animal life. (Biology and life are crucial here.) So a whole dimension of appreciation, so central to our interest in humanly produced works of art, would be absent in the case of computer-produced works, which we ought therefore to view as more akin to crystals than to drawings or paintings. As Frank Palmer puts it, one of the things we value in works of art is 'our negotiation with a human mind other than our own'.[17] And in this negotiation we simultaneously extend our experience and at times find corroboration for our own feelings and viewpoints. At other times, we doubtless find challenges to our own feelings and viewpoints, but the challenge is a challenge precisely because it is a human being—a fellow-traveller on the journey through life—who sees and feels differently from us. Have we missed something? Are we hiding something from ourselves? Or are we dealing with a moral or aesthetic idiot?

In the passage about Mantegna and Veronese from which I have already quoted, Ruskin says of the great artist's power and control

> determine for yourself whether a manhood like that is consistent with any viciousness of soul, with any mean anxiety, with any gnawing lust, any wretchedness of spite or remorse, any consciousness of rebellion against law of God or man, or any actual, though unconscious violation of even the least law to which obedience is essential for the glory of life and the pleasing of its Giver.[18]

Ruskin's attitude is surely preferable to the prevailing attitude of our time, to degrade any hero by any means possible, but it is unsustainable as it stands. Not only are many great artists on a human level mean, spiteful and even worse; some have made it their aim to present human life as mean, spiteful or worse.

[17] F. Palmer, *Literature & Moral Understanding* (Oxford, 1992), p. 159.

[18] Ruskin, *Lectures on Art*, Lecture III, Section 71.

Nevertheless, as always when he is most obviously wrong on the surface, there is something Ruskin is hinting at, something true and important, something which should lead us to take a Bacon or a Céline or a Picasso seriously, even while rejecting the morality of what they have to say and averting our eyes decorously from the sordid details of their biography. Art is closely linked to fantasy. Most people's fantasies are second-hand and often imprecise, rhetorical and incomplete to boot. Many works of art simply mimic other works of art, other people's fantasies in other words, without resting on any basis of skill or observation. The danger of this happening is particularly acute in the mechanistic arts, such as film and photography, where over to the nature of the medium involved a verismilitudinous gloss is inevitably given to any material, even to the most shallow. Where in a novel, dialogue and characterization of the level to be found in the average film or television drama would immediately reveal its unreality, on screen the mechanism involved ensures a semblance of an unbroken and unquestionable truth. By contrast, an imperfectly executed or observed drawing or painting is easily seen through, both for its lack of skill and its absence of originality.

In Ruskin's terms, though, Picasso and Bacon would rank as powerful artists: men with new and direct intentions, and the skill and exertion to convey them in visual form. Analogously, in the field of literature, few writers this century come anywhere near Céline in their ability to re-create a whole world and its meaning, a harsh and vexatious world to be sure, but to do so with neither remorse nor sentimentality.

The visions conveyed by Picasso, Bacon and Céline are neither unworked nor second-hand, and if, in the end, we find them transgressive or ignoble, this is not because they are unrooted in intelligent encounters with human reality. Rather they emphasize aspects of human reality we might prefer to overlook or feel need to be counterposed with other perspectives. Unlike pure fantasy, they are true to the extent that they demand a considered response from those who find them uncomfortable, and not just the instinctive rejection of pornography or of a Nazi tract.

Works of art, then, are human creations, made with skill and craft to evoke and express human meanings. They are also and characteristically singular objects, unique in themselves and reflective of one person's intelligence, sensitivity and skill. Even if a work of art is reproducible, it cannot be machine-generated, for that will be to undermine the role of the artist and the role of work of art as something intended as such by another human being. Equally, its appearance should not be too slick, belying its origins

155

in the struggle of an imperfect, but free human being with his imperfect material: too much ostensible perfection may simply reflect an attempt to deny human freedom, spontaneity and creativity. Art which aims at complete suppression of human freedom and a complete hiding of the artist's personality is, to that extent, an inhumane and imperfect art, even, or perhaps especially where the motivation for such suppression is religious or other-worldly. Beautiful as such art may sometimes be, it may at the same time witness to an attempt by the society from which it derives to suppress and deny human autonomy, and to treat artists as means rather than as ends. To the extent that this is evident in the art, we, with our respect for human autonomy, are bound to find the art less than complete.

Such, indeed, was Ruskin's view:

> Go forth again to gaze upon the old cathedral front where you have smiled so often at the fantastic ignorance of the old sculptors: examine once more those ugly goblins and formless monsters, and stern statues, anatomiless and rigid; but do not mock at them, for they are signs of the life and liberty of every workman who struck the stone; a freedom of thought, and rank in scale of being such as no laws, no charters, no charities can secure; but which it must be the first aim of Europe at this day to regain for her children.[19]

Great art, in Ruskin's view, is never perfect, or finished like machine work. We, as human beings, are fallen and multiply imperfect; it is the aim of mass production, of the division of labour and of technological aids to production to occlude this fact. The perfections afforded by technology are degrading in comparison to the old crafts precisely because they *are* perfect: 'it is only for God to create without toil; that which man can create without toil is worthless', it allows no room for any creative or expressive lapse.

Part of what is at issue here is an aesthetic question: Michelangelo's David or the Rondanini Pietà? Michelangelo as a whole or Canova? Raphael or Leonardo? The Hermes of Praxiteles or the Anavyssos Kouros? Ruskin's own preference for the Gothic over the Classical and particularly over the neo-Classical is well known, though his arguments for the preference are far from convincing. Apart from anything else, he underestimates the freedom for invention and ingenuity which the classical orders permit, while at the same time over—estimating the extent to which the twelfth- and thirteenth-century cathedral builders were innocent of specialism, division of labour and technological advances.

[19] John Ruskin, *The Stones of Venice* (1853), Vol. 2, Ch. VI, Section 15.

At the same time, Ruskin's idealized and unfair characterization of the Gothic does bear on the question of technology in two ways, one of direct relevance to the artistic, and one of wider import. One besetting artistic sin is dominance by too easy a technique to avoid the business of hard creativity, creativity against a background of rules and resistant material which, by curbing fantasy liberates true thought. The salon sculpture and academic painting of the last century was produced by men who had mastered much of the technique of, say, Canova or David, but who used their technical prowess to conceal the fact that they had nothing really to say, and were simply repeating old clichés. As Reynolds put it in his *Twelfth Discourse to the Royal Academy*,

> a provision of endless apparatus, a bustle of infinite enquiry and research ... may be employed to evade and shuffle off real labour—the real labour of thinking[20]

(words which have application, I think, in areas nearer home). To be sure, the Friths, the Makarts and the Geromes and the Gibsons bustled and laboured, and filled rooms full of sculpture and walls full of canvass, but one feels a slackness in the work and an emptiness in the conception—faults, if anything, made worse by their technical facility. One feels an avoidance of a true engagement with the medium, with its difficulties and hence with its possibilities.

The danger of meretriciousness, of the avoidance of creative engagement with the medium are clearly the greater the more the artist is working through technology, with a medium which does much of the work for him. Looking at the visual and musical material we are surrounded with for so much of our lives in 1995, it is hard not to feel the force of Goethe's dictum that technology in alliance with bad taste is the enemy of art most to be feared, and hard, too, not to conclude that the very ease technology affords to fantasy and imagination may not be past of the cause of our artistic decline.

The wider import of these reflections on art and technology is simply to ask whether there might not be something important missing from a life or a culture in which technical proficiency dominates over rather considerations. We have a culture or, perhaps better, a civilization of great technical proficiency, one in which the steam-hammers of the last century are now museum pieces not without their own beauty. Our society is not based on some hellish Nibelheim, but is one in which dreams of power and glory have, for the most part, been replaced by a universal aspira-

---

[20] Joshua Reynolds, *Twelfth Discourse to the Royal Academy* (1789).

tion for comfort and entertainment. We do not have the division of labour, as envisaged by Ruskin—whose result is men divided and broken so as to be capable of one thing only—so much as the abolition of *labour* and its replacement by a huge middle class of administrators and facilitators. Indeed, I see the most accurate portrait of *our* age as painted not by Ruskin, but by de Tocqueville:

> Above this race of man stands an immense and tulerary power, which takes it upon itself alone to secure their gratifications and to watch over their fate. That power is absolute, minute regular provident and mild. It would be like the authority of a parent if, like that authority, its object was to prepare men for manhood; but it seeks on the contrary, to keep them in perpetual childhood: it is well content that people should rejoice provided that they think of nothing but rejoicing...[21]

There is much that could be said about de Tocqueville's vision, and much of that is not directly relevant to our topic. I would, indeed, distance myself from the hint of big brother—of conspiracy—of a *single* tutelary power—in what is supposed to be a description of democracy. But if, as I believe, we are in a situation in which for many if not most people, basic needs are satisfied without much effort of labour, this is because of the advance of technology. At the same time, though, technology infantilizes, encouraging people to be satisfied with the material delights it makes so easy, and to reduce our sense of freedom and democracy to that of chosing among the delights and 'life-styles' they make possible. As old crafts and skills, and the apprenticeships they required, become redundant, there is an increasing desire for a certain type of perfection, but of a bland, mass-produced, unindividualistic type, unconducive to the labour, risk and insight on which true art depends.

The question this leaves us with is whether the ease in living and in entertainment made possible by contemporary technology is contributing with us to an absence of the skill and taste on which good architecture and good art depends.

---

[21] A de Tocqueville, *Democracy in America*, Vol. II (1840), Pt IV, Ch. 6.

Tools, Machines and Marvels

STEPHEN R. L. CLARK

1. Masterful Machinery

Technology, according to Derry and Williams's *Short History*, 'comprises all that bewilderingly varied body of knowledge and devices by which man progressively masters his natural environment'.[1] Their casual, and unconscious, sexism is not unrelated to my present topic. Women enter the story as spinners, burden bearers and, at long last, typists. 'The tying of a bundle on the back or the dragging of it along upon the outspread twigs of a convenient branch are contributions [and by implication the only contributions] to technology which probably had a feminine origin'.[2] Everything else was done by *men*, and what they did was *master*, *conquer*, and *control*. It is also significant that Derry and Williams take it for granted that 'the men [sic] of the Old Stone Age, few and scattered, developed little to help them to conquer their environment':[3] until the advent of agriculture, and settled civilization, there was, they say, neither leisure nor surplus. Later investigation strongly suggests, on the contrary, that hunter-gatherers have time to play with, and keep their 'surplus' where it belongs—out in the world. Settled civilization brought us—or most of us—discipline, time-keeping and back-breaking labour. Maybe, as they suggest, Sumerian priests were 'the first leisured class',[4] but the emphasis should be on 'class'. In more egalitarian, less 'civilized' days, we were all as leisured as any other primate group, and doubtless as preoccupied with mutual grooming, mutual manipulation and the decorative arts.[5]

Back in those old days any competent adult could construct the tools and other devices she might need: digging sticks, string bags,

[1] T. K. Derry and T. I. Williams, *A Short History of Technology*, (Oxford: Clarendon Press, 1960), p. 3.

[2] Ibid., p. 190. Cf. L. Mumford, *The Myth of the Machine* (London: Secker & Warburg, 1967), pp. 140f: 'As home-maker, house-keeper, fire-tender, pot-moulder, garden-cultivator, woman was responsible for the large collection of utensils and utilities that mark neolithic technics: inventions quite as essential for the development of a higher culture as any later machines.'

[3] Ibid. p. 3.

[4] Ibid. p. 7.

[5] See M. Sahlins, *Stone Age Economics* (London: Tavistock Press, 1972).

cooking pots, baby slings, feather ornaments and finger paints. Maybe (most probably) some people made them better, but the crafts were all familiar ones, and could be commemorated in song and story by people who knew what they were talking about. As the crafts grew ever more refined and specialized (as well as ever more vital to our lives) the singers and talkers talked less and less about them. Homer could interest his audience in the making of a shield, but even his main topic was the glory and the pain of war, and the troubles of the human heart. When Virgil tried to write didactic poems about farming, it may be doubted that real farmers even paused to say 'And how does Virgil know?', because no-one would ever have thought he did. Farmers, of course, were reckoned the mainstay of the republic, and their craft accordingly respectable. Crafts that are concerned with the making and mending of less worthy things have not retained the respect that, maybe, smiths once had. Such things, to judge from literature, matter much less than family feuds and love affairs. 'Serious' literature must be about the chattering classes and their problems. Even when writers turn to talk about the people who make and run the machines (rather than those who merely use them) it is on the understanding that their craft is less interesting, less important than their characters, and those latter only insofar as they are cultured. Literature that instead reminds us of the natural laws by which we live, and the crafts that provide us with the setting for those problems, is sure to be despised: weirdly, it is thought to be less 'realistic' than the 'mainstream', 'literary' work.

I am talking, of course, about science fiction, of which Kipling was one great precursor.[6] Tools, machines and marvels, and the possibilities and dangers they create, are the central concern of the one literary genre created in this century[7]—and one largely neglected or misrepresented by its literary critics. Kipling loved them, and loved to give them voices:

We can pull and haul and push and lift and drive,
We can print and plough and weave and heat and light,
We can run and jump and swim and fly and dive,
We can see and hear and count and read and write!...

[6] See S. R. L. Clark, 'Alien dreams', in D. Seed (ed.) *Anticipations: Early Science Fiction and it's Precursors* (Liverpool University Press, 1995).

[7] See G. Westfahl, 'On the true history of science fiction', *Foundation*, **47** (1989), and '"An idea of significant import": Hugo Gernsback's theory of science fiction', *Foundation* **48** (1990), pp. 26–49: science fiction has many precursors, from *The Odyssey* to *Frankenstein* and *The Time Machine*, but has achieved *genre* status only in Hugo Gernsback's time.

But remember, please, the Law by which we live,
We are not built to comprehend a lie,
We can neither love nor pity nor forgive,
If you make a slip in handling us you die!...
Though our smoke may hide the Heavens from your eyes,
It will vanish and the stars will shine again,
Because, for all our power and weight and size,
We are nothing more than children of your brain![8]

His fascination, and his caution, are echoed in science fiction, and that verse marks some of the issues I shall be exploring here: in particular, it gives the lie to the notion that it is by machinery that 'we control the environment'. Rather, machines now constitute a major part of our environment, and are not easily controlled. 'Control' is a part of what Spengler identified as Faustian technics, the expression of Western Culture's urge to master and at last transcend the given. Magian technics, which he located in a distinct Magian or Arabian Culture, is at once more playful and more realistic.[9]

The Faustian attitude (to continue with Spengler's label) is manifested in two contrary forms. According to the first, manipulative engineers are dedicated, pure, self-sacrificing. According to the second, they are knaves, dedicated only to the overthrow of 'natural balances'. Engineers, we are constantly instructed, ought to be guided by a strong professional ethic. Florman's contrary suggestion, that 'what society needs is not more morality from its engineers, but less',[10] seems wilfully paradoxical. But he is surely right to point out that whatever damage our great engineering projects have done, they were initiated with high hopes, and firm resolves. Gibson's short story, 'The Gernsback Continuum', imagines what future might have been, if the founding father of the genre, Hugo Gernsback, had had his way: 'it had all the sinister fruitiness of Hitler Youth propaganda'.[11] Engineers would transform the world, and human nature. Fashionably environmentalist distaste for the results is merely the mirror-image of that project, and often just as

[8] *Rudyard Kipling's Verse: Inclusive Edition 1885–1926* (London: Hodder and Stoughton, 1927), p. 675.
[9] See O. Spengler, *The Decline of the West: II Perspectives of World History*, tr. C. F. Atkinson (New York: Alfred A. Knopf, 1928), pp. 233 ff.
[10] S. C. Florman, *The Existential Pleasures of Engineering* (London: Barrie and Jenkins, 1976), p. 38.
[11] W. Gibson. 'The Gernsback Continuum', in *Burning Chrome* (London: Grafton Books, 1988), pp. 36–50 (first published 1980). Strictly, Gernsback may be misjudged: he had more doubt about the technocratic future than Gibson may have supposed: see Westfahl's 'An idea of significant import'. (See note 7).

megalomaniac, just as dedicated to remaking the world, and human nature. What we need is not 'morality' (in that sense), but an awareness of the limits to morality and human power, and of the self-deceiving tendencies of human moralism.

Much of the opposition to 'ironfounders and others', after all, is snobbish, nationalistic, or entirely hypocritical. The anonymous Welshman who lamented the cutting of Glyn Cynon cursed the invading Saxons who had destroyed his land of happy memory.

> Many a birch tree green of cloak
> (I'd like to choke the Saxon)
> Is now a flaming heap of fire
> Where iron-workers blacken.[12]

But he had his own uses for the land, and his own beloved machines. When a later Englishman made a similar complaint, he too made some distinctions:

> Your worship is your furnaces,
> Which like old idols, lost obscenes,
> Have molten bowels; your vision is
> Machines for making more machines.[13]

But when the end has come:

> The middens of your burning beasts
> Shall be raked over till they yield
> Last priceless slags for fashionings high,
> Ploughs to wake grass in every field,
> Chisels men's hands to magnify.

The machinery one generation thinks is dreadful innovation will invite nostalgic affection in a later—and very few of us take steps not to profit from the ironworkers' actions. Many childrens' books of the 1970s chose to emphasize the damage that industrialists have done: the mad professor's marvellous machine that does nothing very much but grow, and spew out gunk; the world abandoned as a waste dump till the ecologically minded dinosaurs wake from sleep and clean it up. Stories of the 1980s and 1990s seem to be about a different escape, into the dreaming computers. Which is perhaps some confirmation of Spengler's prophecy: it is not that we shall 'run out of material', but that 'the West European-

[12] G. Jones (ed.), *The Oxford Book of Welsh Verse in English*, (Oxford: Clarendon Press, 1977), p. 71; see S. R. L. Clark, *Civil Peace and Sacred Order: Limits & Renewals Vol. I* (Oxford: Clarendon Press, 1989), pp. 86–87.

[13] Gordon Bottomley, 'To Ironfounders and Others', In Quiller-Couch (ed.), *The Oxford Book of English Verse* (Oxford: Clarendon Press, 1939), p. 1109.

American technics *will itself have ended* long before'.[14] The coming technics, I shall suggest, is Magian.

'The machine-technics will end with the Faustian civilization and one day will lie in fragments, *forgotten*—our railways and steamships as dead as the Roman roads and the Chinese wall, our giant cities and skyscrapers in ruins like old Memphis and Babylon.'[15] But this will only be a necessary pause before another technics takes shape. One reason—amongst many—why Spengler is despised is that he 'would sooner have the fine mind-begotten forms of a fast steamer, a steel structure, a precision-lathe, the subtlety and elegance of many chemical and optical processes, than all the pickings and stealings of present day "arts and crafts", architecture and painting included'.[16] Control is one thing: elegance another.

In this dangerous speculation? Any Civilization, in Spengler's sense, is at war with Life, and the Faustian (which seeks transcendence) more than most—but we ought now to be wary of the attempt to reawaken Life.

All things organic are dying in the grip of organization. An artificial world is permeating and poisoning the natural. The Civilization itself has become a machine that does, or tries to do, everything in mechanical fashion. We think only in horse-power now; we cannot look at a waterfall without mentally turning it into electric power; we cannot survey a countryside full of pasturing cattle without thinking of its exploitation as a source of meat-supply; we cannot look at the beautiful old handwork of an unspoilt primitive people without wishing to replace it by a modern technical process.[17]

If the Nazis had won they would be claiming to be on the side of Life. 'There is no inorganic nature, there is no dead, mechanical earth. The Great Mother has been won back to life', said one of their ideologues.[18] Because they lost (as Spengler, of course,

---

[14] O. Spengler, *Man and Technics*, tr. C. F. Atkinson (New York: Alfred A. Knopf, 1932), p. 96.

[15] Ibid. p. 103.

[16] O. Spengler, *The Decline of the West: I Form and Actuality*, tr. C. F. Atkinson (New York: Alfred A. Knopf, 1926), pp. 43f.

[17] *Man and Technics*, p. 94.

[18] Ernst Krieck in 1936, quoted in R. A. Pois, *National Socialism and the Religion of Nature* (New York: St Martin's Press, 1986), p. 117. Krieck's feud with Heidegger was no more than a dispute within the Nazi movement: see V. Farias, *Heidegger and Nazism*, tr. P. Burnell and G. R. Ricci, J. Margolis and T. Rockmore (eds.) (Philadelphia: Temple University Press, 1989), pp. 168ff.

expected) we can more easily see that they were themselves a symptom of the mechanizing and manipulative tendency, and of the despair it had engendered. Are we so sure, could Bottomley have been so sure, that we are not?

2. Artificial Lives

'We can see and hear and count and read and write', so Kipling let the machines say, while also denying that they can be anything but children of our brain. Tools and domestic servants, or the idea of these, meet in machines. We have used tools (like digging sticks) from the beginning, and domesticated dogs or slaves or cattle. The first have no wills of their own at all; the second may be praised, or bullied, or seduced into obedience, but have their own agendas. Machines are meant to be obedient, but can do more than tools, which merely magnify our hands. They are the outward images of literal-minded genies, who will do exactly what we ask them, whether we will or no. In the neolithic age any competent adult could make the tools she needed; in the premodern era (from the discovery of metal-working until, almost, now) there were skilled people who could do it, and understood the workings of the machines they made. Kipling again, in McAndrew's Hymn:

> Lord, send a man like Robbie Burns to sing the Song o' Steam!
> To match wi' Scotia's noblest speech yon orchestra sublime
> Whaurto—uplifted like the Just—the tail-rods mark the time,
> The crank-throws give the double-bass, the feed-pump sobs an' heaves,
> An' now the main eccentrics start their quarrel in the sheaves:
> Her time, her own appointed time, the rocking link-head bides,
> Till—hear that note?—the rod's return whings glimmerin' through the guides.
> They're all awa'! True beat, full power, the clangin' chorus goes
> Clear to the tunnel where they sit, my purrin' dynamoes.
> Interdependence absolute, foreseen, ordained, decreed.
> To work, Ye'll note, at ony tilt an' every rate o' speed.[19]

Kipling, rather than any historical reality, probably explains the number of Scottish engineers in science fiction. In these days we are at the brink of an era when no-one really understands the machines we tend and pray to. They are fast reaching the moment when they rival living forms in their complexity. We can no longer

[19] Kipling, *Rudyard Kipling's Verse*, p. 126.

tell what they would do if they were working properly, in accordance with the rules we set them, except by setting them to work. McAndrew knew what his machines should do, and could identify the faults if they did not. The same is still true for some of our machinery: the garage mechanic who can tell at a glance what wire is loose displays a casual intelligence the equal of those professionals who can detect a distributed middle or a begged question. But the machines that are the icons of this age have no such simple fabrics. We do not know what DNA would do if it were left to its own devices in the eucaryotic cells it inhabits, and so cannot tell if it is, or if it is hampered or controlled by invading spirits or morphogenetic fields.[20] We do not know, except by waiting to see, what our computer programs will do, singly or in concert.

Because these new machines, these marvels, are complex beyond the power of our imaginations, and likelier to be understood by guesswork and projective imagery than by mechanical calculation, they are, in a way, like living things, like the creatures we domesticated in the long ago. Even the most marvellous, of course, are more predictable than all but the very simplest products of evolutionary change. Those enthusiasts who, ten years ago, were predicting that we could soon have sensible, human-style conversations with artificial intelligences would now be delighted if we could make a machine to mimic mice. *Artificial Life* is now the catch phrase, and the fantasy is that there will one day be cyberplankton, roaming the Internet to eat up 'viruses' and 'worms'.[21] Maybe, if we could sustain the network for as long, pro rata, as the earth itself, those cyberlife forms could evolve into new rational dreamers like ourselves—and we would no more know how it had happened than we know how we did. You may already be saying that of course we know how we did: natural selection, obviously. I shall return to that delusion in a moment.

Because we don't know now, and will not know hereafter, how these marvels reach conclusions or perform their actions, our postmodern state is rather like our ancestors'. I said that in the long ago any competent adult could make her tools, and that in the premodern era there were at least Skilled Persons who knew what to do. But of course they too did not know *how* it worked. McAndrew's knowledge is anomalous. For a very brief period we could entertain the thought that we could see, or he could see, exactly what was happening at the microlevel to produce the effects he wanted. The great smiths of our past could suffer no

[20] Which is why 'genetic engineering' is so hit-and-miss: we rarely know what bit of DNA is responsible, in general, for any particular phenotypic feature, and never know exactly how it is.

such illusion. 'Cementation' was apparently invented in the Hittite Empire in about 1400 BC: 'this was a process for steeling wrought iron bars by repeated hammerings and heatings in direct contact with charcoal, which diffuses carbon into the surface regions of the metal'.[22] Tempering, by quenching the hot metal in cold water, seems to have originated two centuries later. The process worked, but no-one need know why.

Such tools, of war or agriculture, had specific purposes within the culture of their day. They were, that is, something more than merely bars of iron 'Every machine *serves* some one process and owes its existence to *thought about this process*.'[23] So in making them, by chance discovery or baffling intuition, the great smiths channeled 'spirits'. Some of them, perhaps, imagined that they were releasing, or moulding, spirits resident in the ore. Others perhaps imagined that they were invoking spirits from another place. Whatever uncomprehended power it was that moulded animal flesh into a living form, could also (through the smiths) mould metal. Why might it not create all-metal automata, as Hephaistos, god of smiths, is said to have done? Why should the spirit that changed iron ore into a unitary, well-forged sword not also turn it, in the hands of a warrior, into a living being? Who could tell what would be possible? Not knowing how the effects are achieved, but only that they are, we can place no limits on what other mysteries could open up. Finding that the products of the art have their own life and meaning, we can imagine their production as an invocation, or conception. Only in the few industrial centuries can we easily imagine that we understand the processes, and seek to exclude all reference to imagined spirits. Consider Thomas Sprat:[24]

> The poets of old to make all things look more venerable than they were, devised a thousand false Chimeras; on every Field, River, Grove and Cave they bestowed a Fantasm of their own making: With these they amazed the world ... And in the mod-

[21] There is already a minor, roving program ('the cancelbot') designed to eliminate all e-mail from a law firm which offended the older inhabitants of Internet by mass-mailing an advertisement: see *Time* 144.4, p. 52 (25 July 1994). It may be, of course, that such a program, however devious, will no more be 'alive' than the computer simulation of a thunderstorm is wet—but I am not sure that we can be sure of this.

[22] Derry and Williams, *A Short History of Technology*, p. 121.

[23] Spengler, *Man and Technics*, p. 11.

[24] *History of the Royal Society* (1702, p. 340) vs fairies (cited by B. Wiley, *The Seventeenth Century Background* (London: Chatto & Windus, 1934), p. 213).

ern Ages these Fantastical Forms were reviv'd and possessed Christendom ... All which abuses if those acute Philosophers did not promote, yet they were never able to overcome; nay, not even so much as King Oberon and his invisible Army. But from the time in which the Real Philosophy has appear'd there is scarce any whisper remaining of such horrors ... The cours of things goes quietly along, in its own true channel of Natural Causes and Effects. For this we are beholden to experiments; which though they have not yet completed the discovery of the true world, yet they have already vanquished those wild inhabitants of the false world, that us'd to astonish the minds of men.

The ironic conclusion of Sprat's Real Philosophy has been the production of machinery on which we can, with far more plausibility, bestow our Fantasms. Not knowing how they work, and not being able to predict what they will do by mechanistic reasoning, we are at liberty to imagine that they feel themselves the purposes we made for them.

Clarke's Axiom (Arthur C. Clarke, that is) is that any sufficiently advanced technology is indistinguishable from magic. All that he meant, I suspect, is that there are no upper limits on what we can learn to do with the help of Sprat's Real Philosophy: what could only be imagined can now be performed, and will look 'like magic' to the uninitiated. It is also significant that the new technology allows us to achieve results by words: we *tell* computers what to do, and our words have real effects. That was always the essence of magic: to extend our verbal control of things. It follows that we are increasingly surrounded by entities that may be essentially mechanical, but that must be dealt with as if they were semantic. At the same time we can entertain the fantasy of 'virtual reality', of making illusions so exact, so convincing that they allow us to forget the material base from which they rise.

This projection of our purposes, these Fantasms, is not only observed in popular fiction and idle conversation. It is not just Kipling and his descendants who imagine voices for machinery, nor only irritated motorists who speak of their car as 'feeling fractious this morning'. Respectable philosophers are ready to conceive that the new machines will be like living things, and living things, by parity of argument, like new machines. 'Really' they work mechanically, but no-one can imagine how, and therefore we impute to them, as causally effective powers, the purposes they serve in human life. The imputation is harmless enough, as long as we know that it isn't really true. Once upon a time, ghosts, memories and secondary qualities were only projections of the human mind; now that mind turns out to be a projection too.[25] Once we were warned against the pathetic fallacy that attributed motives or feelings to nature, or to beasts; now we are warned—bizarrely— against attributing them to our very selves.

That warning, delivered by 'eliminative materialists', comes in two flavours. The weaker merely predicts that we might find a different range of metaphors for the inner life that, undoubtedly, we have, and that some enlightenment assumptions about the transparency and unity of 'the human mind' are mistaken. It is difficult to deny that this could be true, and difficult not to suspect that this is no real elimination. The stronger flavour suggests that 'consciousness', 'qualia', 'intentionality' are either as unreal as phlogiston or else to be understood merely 'objectively'. A machine that responds to external stimuli in accordance with some definable rule is held to be as 'conscious', as 'alert', as anything at all could be. I find it difficult to understand how anyone not blinded by a theory could possibly take this seriously. The blindness is explicable as the effect of convergent intellectual tendencies: objectivism, mechanism and the denial of 'the given'.

All this is an echo of the Magian. Whereas Faustian stories deal with the rise, or fall, of solitary and autarchic heroes, the other tradition acknowledges the importance of quick wits and courage, but envisages its temporary heroes as non-autarchic parts of a wider world.

> Whereas the Faustian man is an 'I' that in the last resort draws its own conclusions about the Infinite; whereas the Apollinian man, as one *soma* among many, represents only himself; the Magian man, with his spiritual kind of being, is only *a part of a pneumatic 'We' that, descending from above, is one and the same in all believers.*[26]

Will, which is crucial for the Faustian, is irrelevant for Magians. 'The idea of individual wills is simply meaningless [in Magian thought], for 'will' and 'thought' in man are not prime, but already effects of the deity upon him'.[27] In the Arabian Nights there are mechanical horses that can do as well as living ones, and people who can be turned into statues. The stories are told, remember, to a control-freak who has his wife-of-the-night executed rather than

[25] H. Putnam, *The Many Faces of Realism* (La Salle, Illinois: Open Court, 1987), p. 15, comments on the absurdity; see also, S. R. L. Clark, 'Minds, memes and rhetoric', *Inquiry*, **33** (1993), pp. 3–16.

[26] Spengler, *The Decline of the West II: Perspectives of World History*, p. 235: Plato, by this criterion, was not clearly Apollinian.

[27] Ibid. pp. 235f.

risk her infidelity. What they gradually reveal to him is the absurdity of his desire, the hugeness of the world he hoped to rule. Everything, in one way, behaves according to the laws that govern Kipling's engines, and in another operates within a moral universe where there is never only one hero, only one centre of attention. To avoid misunderstanding: Spengler attributes these attitudes to a distinct non-Western culture; my view is that they are simply another strand in human thought. The Magian Consensus, in particular, is not territorial.[28] Really these terms name tendencies, not discrete Cultures, and many Great Names of the Western Tradition turn out to have been Magians (e.g. Augustine, Spinoza).

3. Philosophical Fantasms

Objectivism, so to call it, is the Enlightenment project of identifying primary reality only with what can be understood without reference to any particular perspective or evaluation. 'Objective' facts, by definition, are ones that include no reference to moral or other value, nor to any subjectively supported sensa. It follows at once that such objective facts cannot 'explain' those evaluations or perspectives. Postulating the 'objective' world we create a weird problem for ourselves: why are there qualia, or subjective intentions, at all? The problem is admirably expressed by Joseph Glanvill, a fervent Cartesian, in his *Vanity of Dogmatizing*, in the course of an enumeration of the limitations of human knowledge. 'How the purer Spirit', he writes, 'is united to this Clod, is a knot too hard for fallen Humanity to untie. How should a thought be united to a marble statue, or a sun-beam to a lump of clay! The freezing of the words in the air in northern climes is as conceivable, as this strange union. ... And to hang weights on the wings of the winde seems far more intelligible'.[29] The problem is not, as some suppose, that Descartes arbitrarily invents a mind distinct from matter. The point is that science rests on the construction (the hypothesis) of a world where wishes are *not* horses, where things are *not* the case merely because we say they are. To admit that they sometimes are is to admit that 'an empty wish should remove Mountains', as Glanvill puts it. To deny that they ever are seems to leave no room for efficacious volition. Objectivism has for long been saved from abject absurdity by speaking of 'emergent' properties, not contained in the original reality but somehow produced 'by magic'. But this is an idea as suspect as Leibniz said:

[28] Ibid. p. 320.

[29] Wiley, *The Seventeenth Century Background*, p. 84, quoting Glanvill (1661, p. 20).

Stephen R. L. Clark

an appeal to over-occult properties and inexplicable faculties ('helpful goblins which come forward like gods on the stage, or like the fairies in *Amadis*, to do on demand anything that a philosopher wants of them'[30]). Despite attempts to find analogies for such an emergence in such concepts as 'liquidity' (a property of many molecules of water apparently not shared by individual molecules, or constituent atoms), the truth is that there are no cases of this supposed event except ones constituted by the very relationship of 'objective' and 'qualitative' being that concerns us. Liquidity, as an objective property, bears a precise and methematically describable relationship to the molecules of water: only as a qualitative property does it pose a problem—but that is the problem.

So it is not surprising that investigators convinced that reality is 'objective' (and that the best mode of investigation requires a studied 'objectivity', self-alienated from any 'emotional' or 'perspectival' gaze—which is another version of the Faustian drive) will seek to deny reality to the very thing that their predecessors identified as the—really existing—realm of illusion. In doing so they agree with another modern or postmodern strand of thought: the denial of 'the given'. Whereas theorists used to distinguish theory and data (and hope for theories that adequately accommodated data), it is now more fashionable to deny that there can ever be pre-theoretical data. Eliminativists will therefore insist that immediate sense (including introspection) is so infected by outdated theories as to require immediate disinfection. The weaker flavour, as before, means only that our perspectival and qualitative experience may be transformed by adopting a new theory (a notion which anyone could accept). The stronger (and less palatable) flavour apparently denies that anything at all is 'data'—and there is therefore nothing that their theories should accommodate. Opponents (such as myself) conclude that this is to surpass the supposed excesses of scholasticism from which the Enlightenment, perhaps, once rescued us. Both objectivists and non-objectivists, however, may end—in practice—by rejecting the Cartesian division between a merely objective and a qualitative world. The world we must deal with is full, metaphorically or really, with the Fantasms Sprat renounced.

4. The Coming Age?

Eliminativism, of either flavour, would not be so fashionable were it not for mechanism, the underlying notion that engineers—a class that here includes pharmacists and surgeons—can create

[30] G. Leibniz, *New Essays on Human Understanding*, P. Remnant and J. Bennett (eds) (Cambridge University Press, 1981), p. 382 (4.3.7).

what moralists and mystics only dreamt about. Maybe Spengler was right to see in this the effort of a non-creative Civilization to repeat, surpass and revenge itself upon an earlier, inspired Culture. But in its origins, and even in the form that McAndrew gives it, mechanism is a decently humble and humanistic project. Moralists and mystics—or Faustian ones—have tended to exalt the human will: if only we wish for something strongly and single-mindedly we shall win the ball-game, pass the exam, purify the human heart, move mountains and rise up to heaven. How could we lose when we are so *sincere*? Those who fail to do these things have only themselves to blame (and if they fail to do so, we will do it for them). On this account the engineering spirit is not a triumph of the human will, but a proper recognition of the will's limitations. Wishes are not horses: if we wish to ride we have to work alongside things as they are, accepting guidance and preferring small successes. Fortunately for us all, God does not reward the just nor punish the unjust. Yogis who spend their lives learning to levitate across a river that can be crossed, by ferry, for a copper coin are wasting time. Telling the clinically depressed to pull themselves together is in all ways less successful than such simple, engineering tricks as giving them more light, more exercise, more B-vitamins, less smog. Truly saintly people recognize that they could easily have done (in a sense, almost have done) whatever evil act moralizing journalists and judges currently condemn. Better change wrong-doers' circumstances than condemn them. Better turn to engineers, subtly aware of the Laws by which we live, than imagine we can *will* ourselves to heaven.

Consider, though, a classic work of science fiction, *The Humanoids*. Ten thousand years from now, in a human species scattered across the local cluster, an engineer, losing hope in human decency, devises self-reproducing robots, the humanoids, linked in an interstellar network, with the single command 'to serve and obey and guard men from harm'. Most of the novel concerns attempts to resist the absolute control the humanoids impose, a control that culminates in the transformation of humans into limbs and organs of the interstellar network, all in the name of 'total happiness'. In the concluding pages the last resistance fails, and comes to seem absurd: what is there to fear in a system that protects us from the natural consequences of our own ignorance, folly and vice? The novel's central character, remodelled by the network, is put in charge of further extensions:

> Forester wondered why his body tried to stiffen, and why he almost shook his head. He could recall a time when he had dis-

171

liked the humanoids and even mistrusted Frank Ironsmith [their human ally, his former subordinate], but now, even though his recollection of past events seemed clear enough, all the misguided emotions which must have driven him to his unfortunate past actions were fading from awareness, even as he fumbled vaguely for them, like the irrelevant stuff of some unlikely dream.[31]

And why not? The beliefs and emotions Forester feels at the end are engendered by direct manipulation of his brain and hormones. Would they have been more 'authentic' if they had been induced by ordinarily rhetorical persuasion, or were a 'natural' outgrowth of random neural and hormonal events, feebly configured by the accidental norms of his immediate society? Ironsmith is convinced throughout that he has access to a rational order, and that the humanoids will serve that goal. Forester is equally convinced, at first, that the humanoids are enemies: but what reason could there be to reckon either impulse any truer than the other? All that can sensibly be said, outside the story, is that the humanoids win. That victory is made more bearable, incidentally, because they begin to take psychokinetic power seriously (which is, magic).

Neither victims nor manipulators, in such stories, are 'human' in the sense that most of us imagine. The manipulators know how different ideals and attachments can be made, in others and in their own originals. In C. J. Cherryh's *Cyteen*, perhaps the most brilliant of recent psycho-historical fantasies,[32] the imagined psychologists who rule a future human society, called Union, have begun to understand human motivation so well (after prolonged experimentation with extra-uterine conceptions, drugs, tape-learning) that they can both manipulate whole populations, and hope to replicate whichever of their number dies untimely. In Cherryh's universe there is a natural and legal distinction between tape-taught azi and hand-reared humans—but the real distinction rests only in their relative susceptibility to instruction. The hand-reared's relative immunity is a product of their exposure to multiple, apparently random stimuli, so that they are never wholly convinced of anything. That fact itself makes them manipulable, in other ways, by those who understand. Some reviews of the book complain that Cherryh's characters sometimes 'act out of character'. It would be more accurate to say that her characters, though admirably characterized for literary purposes, do not have characters at all, as we might understand them. Knowing that their own

---

[31] J. Williamson, *The Humanoids* (New York: Lancer Books, 1963; first published by Simon and Schuster, 1949).

[32] C. J. Cherryh, *Cyteen* (London: Hodder and Stoughton, 1988).

ideals and attachments are products, they can more easily dispense with them, as they see fit. They may 'believe' (at some level) that they are working for 'human welfare' over the long run (like Ironsmith), but they can take this seriously only by inattention. In some of Cherryh's other stories species disloyalty is almost a way of life, and the idea that there could be 'objective' criteria for 'better outcomes' is recognizably absurd. Cyteen itself, in other stories, is the enemy of a more recognizably 'liberal' culture—which is just as much a product of its own history and modes of trade and manufacture.

In the coming age, we have imagined, there will be no more 'wild humans'. None of us, of course, have been entirely 'wild' or 'savage' since before our species first began. Certainly the occasional approving or disapproving fantasy of free-born individualists is simply fantasy.[33] But humans have in a way been wild: very few societies have been able to enforce one way of living, one consistent set of assumptions, or avoid the chance development of counter-ideas and feelings. Wheat and tares grow in the same field till the harvest. Domestication, selective breeding, and controlled environments have done much to weed the fields and lineages of our crops and cattle. We have also domesticated, bred and sought to control ourselves. In so far as we can model our own behaviour through our new marvellous machines, or through millions of experimental humanoids, or azi, we can dream that our agricultural success will be followed by a social one. Treating each other as machines to be mended or remade is surely better than treating each other as morally feeble people to be inefficiently reprogrammed, or as evil swarms to be savaged. It doesn't follow that there is or could be *one* controller: rather the opposite.

The truth is: control is impossible. None of us know the effects of anything we do: in other words, we do not know what it is we are choosing to do, nor why we choose. I do not even know what it is I have succeeded in conveying by these words. We can perhaps have small successes: the larger our ambitions, the less chance we have of fulfilling them. Which is what the Magian strand of our culture told us all along.

5. Natural Selection

Our successes will, at best, be local ones, and so is our knowledge. Natural selection does not guarantee, or even make it likely that

[33] See S. R. L. Clark, 'Good ethology and the decent *Polis*', in A. Loizou and H. Lesser (eds), *The Good of Community*, (Aldershot: Gower Press, 1988).

creatures like ourselves will know the Whole Truth plain. To that extent the whole grand theory of our biological evolution is deeply contradictory. We have the biological markers, the *Umwelt*, the theories and deep convictions that we do because we have evolved to have them. It isn't even true, on current theory, that the ones we have have 'won' a real contest with alternatives. They simply haven't lost (like scientific theories on Popper's doctrine).[34] It may seem obvious, accordingly, that our natural impulses, moral and epistemic, have no authority, and might as well be changed to fit an engineered, and fully domestic system. Unfortunately, we have no access to that system. Even if mechanical intelligences arrived from Outside Over There we could have no real faith that they were in any better position than ourselves, being only the children of some other natural brain. But why might we not accept such engineered, or axiomatized perfection, on the plea that—though it is no *truer*—it might, as Protagoras said, be *better*?

Plato's answer, and the one that is hovering unspoken here, is that, after all, we can have access to the Real World, by pursuing Real Beauty in its manifold appearances. On that (Galilean) account, it is also significant that people characteristically use quite different rules for 'natural' and 'artefactual' entities.[35] A coffee pot can be turned into a bird-feeder: no amount of tinkering will turn a racoon into a skunk, or a pawpaw into a pineapple. Because there are real kinds it is peculiarly obscene to treat such organisms as artefacts. I prefer, on this occasion, to invoke another, and more Magian, answer. A fully domesticated, engineered world may be neither more nor less a 'true' or 'just' one. What is almost certainly is, is a less stable and long-lasting one. Monocultures are a very bad, and very shortlived, idea precisely because they are vulnerable to changing circumstances. Variety is not a defect in a natural population. Mutation and difference are not flaws in the reproductive process (as Aristotle let himself believe). They are what sex is for. Cyteen's psycho-historians must take care not to impose a single order on all human populations if they wish humanity to have a future. They must not exclude chance and unpredicted change. By the same token really successful genes, so far from being selfish and unchanging, must allow colossal variation, and act to facilitate other, new developments.

In other imagined futures chaos wins. Our species future does not lie in the fully domesticated, fully engineered condition, but in

[34] See Spengler, *Men and Technics*, p. 82: 'a working hypothesis need not to be "correct", it is only required to be practical'.

[35] See S. Pinker, *The Language Myth* (London: Allen Lane, 1994), pp. 424f.

an efflorescence of distinct possibilities. In Gregory Benford's future history, for example, our descendants survive within the interstices of a mechanized civilization.[36] Like our neolithic ancestors, but at a higher level, they can use artefacts, not wholly knowing how. Their opponents are the Butlerian descendants of such artefacts, operating according to Law, able neither 'to love nor pity nor forgive', and unable to recognize 'real essences'. Those machine intelligences, in the short run, are adaptable and wildly successful. In the long run, their sort of intelligence is wholly inappropriate to a world of change and chance, just as Cherryh's azi, reared to believe in logic, certainty, stability, are wholly unequipped to cope with random, fluctuating change. Azi can, in a way, be 'saints': unmalicious, long-suffering and obedient. But 'saints' like that are not what wild life needs. Cyberpunk, as a fashion within SF, gets some of its support from adolescent angst: but its main merit lies in its deconstruction of the dream of totalitarian, computerized control. Gernsback's paradise is merely another ecological niche, rapidly colonized by creatures doing their own things. The humanoids' network will begin to break up as soon as it is successful. Even Orwell's Newspeak, if it were to be imposed by Party rule, would be 'creolized' into another natural language in a single generation.[37] 'Even if it were possible for our enemies to kill all who are steadfast in the Faith, that killing would be useless to them. For even in their slaves the seeds of Faith would still lie dormant, awaiting the proper hour and the voice of God to flower once more.'[38]

From all of which I conclude that 'technology' need not be Faustian, need not be thought of as a route towards full mastery, as though we could be uncontroversially sure of all our goals and methods, and capable of realizing all and only what we wish. Let us instead accept the other model: the making of occasional aids, of use within a fluctuating, changeful universe that we cannot control. Tools, machines and marvels are all marvellous, and all mysterious. For Magians machines are not devices to extend 'man's power' nor reasons to imagine a merely 'objective' world. They work, as McAndrew realized, according to God's Laws, and even though they are dangerously literal-minded we can work with them.

The world of Magian mankind is filled with a fairy-tale feeling. Devils and evil spirits threaten man; angels and fairies protect him. There are amulets and talismans, mysterious lands, cities,

[36] See G. Benford, *Great Sky River* (New York: Bantam Press, 1987).

[37] See S. Pinker, *The Language Myth*, p. 82.

[38] G. Dickson, *The Final Encyclopedia* (London: Sphere Books, 1985), p. 352.

buildings and beings, secret letters, Solomon's Seal, the Philosopher's Stone.[39]

This adequately describes the world of contemporary science fiction too, and the world at hand. God, or the Unified Theory, is not the sort of thing that we can grasp, though sanity lies in acknowledging Its reality. Perhaps Spengler was right to foresee the end of Faustian technics: he was certainly wrong to think that Magian technics was already dead. There will, he was right to think, be many more technics, of yet unknown varieties; whether any will be a real advance, who knows?

[39] Spengler, *The Decline of the West: II Perspectives on World History*, p. 237; cf. p. 304, which identifies another aspect of fairy-tale with the Faustian outlook, namely the wish to see inside matter.

Values, Means and Ends

ROBERT GRANT

Morals and politics occupy themselves, if not exclusively, then at any rate centrally, with questions of value. Politicians and moralists deplore the alleged decline of values while pressing supposedly new ones upon us. The fiercest sympathies and antipathies, whether between individuals or between societies, are those which stem either from a community or from a divergence of values. 'So natural to mankind,' said Mill, 'is intolerance in whatever they really care about.'

Let us ignore the obvious question, raised in their different ways by both Marxism and sociobiology, as to whether values are ultimately reducible to interests. It might even be a somewhat dull question, since whatever one really cares about must almost by definition constitute an interest. So what, practically speaking, is gained by grounding it in some other kind of interest, when one already has all the information about it one needs? Why should one's caring about a thing require any explanation, so long as it belongs to the class of things commonly recognized, even by Mill, to be worth caring about?

Some more questions: what is it to say that something is valuable, or a value? Is there a difference between a thing's *having* a value and its *being* a value? If I say I prefer *x* to *y*, I am saying that I set a higher value on *x*; but is that quite the same as saying that as far as I personally am concerned *x* constitutes or embodies a value? Can there be such a thing, strictly speaking, as a purely personal value, that is, a value purely *for me*? Would a value, so construed, be anything more than a particularly intense preference, something I might see, perhaps, as an object of my personal need rather than merely of my personal desire? If not, would it make any difference if we were to understand values as being objects of intense *collective* preference? If a preference were common to, or, still more, definitive of, a society would that suffice to establish it after all as a value?

I shall try in due course to address these questions, or rather touch on some of them, for to deal with all would require at least a book. I have spelt them out explicitly not because I think I can do them justice (I can't here, and they are daunting enough anyway), but simply in order to stake out the territory, to illustrate in what kind of country our discussion ultimately belongs. It seems to me that they, and the categories in which they deal—value, obviously, but also, as we shall see, means and ends—must be central to any treatment of our common theme, technology.

Some years ago the 'white goods' manufacturer Zanussi used to advertise its products as 'the appliance of science'. (Since 'application' was evidently meant, I take it there was also a feeble allusion, exploited for the sake of the rhyme, to the notion of a 'domestic appliance'.) Zanussi's catchphrase might seem a plausible enough definition of technology, at least to anyone unfamiliar with Lewis Wolpert's book *The Unnatural Nature of Science*.[1] I mention Wolpert here, not because his observations are crucial to my argument, but merely for their intrinsic interest and obvious truth, to the effect that historically most technology, for example that of ancient China, derived from no genuinely scientific knowledge, simply because there was no such thing. However sophisticated it may have been, ancient Chinese technology was fundamentally serendipitous, empirical or *ad hoc*. (Thus much Wolpert.)

We might speculate further. Where new techniques did not just emerge spontaneously or at random, but were actively developed, they must have been developed not by recourse to any theoretical foundations (since there were none), but simply by the extension, direct or analogical as may be, of current successful techniques. If we consider some ordinary examples of the latter (and we need not confine ourselves to China) it is evident that both they and any subsequent spin-off from them need owe nothing to science. The builders and joiners of ancient Egypt knew nothing of Pythagoras, and were probably wholly ignorant of anything like the appropriate geometry. Yet when (as they constantly did) they needed a try-square, they used a 3-4-5 triangle, simply because they had observed that it always contained a right angle.[2] Even today (I am told) there is no physiological explanation of the pharmacological properties of aspirin, a drug derived from a folk remedy (willow bark), yet which technologically speaking must be accounted a brilliant success. Ehrlich is supposed to have discovered Salvarsan, his so-called 'magic bullet' against syphilis, by the patient, purely mechanical screening of hundreds of chemical compounds, most of them dyes. (Salvarsan's alternative name, 606, derives from its being the 606th substance Ehrlich tested.) None of these practical discoveries can really be called scientific. *Per contra*, much genuine science—radio astronomy, say—has no practical application, and is unlikely ever to find one.

[1] London and Boston: Faber and Faber, 1992.

[2] I hope my information is no less reliable for being derived from J. Bronowski's television series of the 1970s, *The Ascent of Man*.

Contrary to common belief, then, technology is not the same thing as applied science, though it clearly overlaps with it. Whether we should rather call it applied *knowledge*, as though the knowledge were prior to its application, or simply see technology as the generic name for various kinds of purposive practice in which, since they succeed in their aim, knowledge must *de facto* be embodied, does not really matter. Two things seem essential to a basic, non-metaphorical, and I hope non-controversial definition of technology.

First, technology proper (I shall come to improper technology later). Technology proper takes (or originally took) as its object the material or physical world. In other words, it is directed towards and upon those things which, since they possess neither self-consciousness nor will, can with both linguistic and moral propriety be called manipulable. (The objects of the *esprit de géométrie*, says Pascal, *se laissent manier*.)[3] To the extent that it has a similar object of attention (though not a similar purpose in respect of it), technology does indeed overlap with science. In Martin Buber's terminology, technology engages not with the world of 'Thou' but with the world of 'It'.[4]

Ideally considered, technology has a human subject but a material object. Its relationship to the world, though intentional, is also mechanical, in that it involves an active 'I' and a passive 'It'. A spider's web is a marvellous, complex, and manifestly functional construction. Nevertheless it would be extravagant, or at best a figure of speech, to call it a piece of technology (as one would, e.g., a fisherman's nets). For, serendipity notwithstanding, technology seems to imply at least some element of ingenuity or deliberate contrivance. It might even be plausible, in the human case, to regard serendipity, or chance technological discovery, as a kind of retrospective invention. In the world of biological nature evolutionarily profitable opportunities arise by chance, but no non-human organism is capable actually of recognizing one when it occurs. If it takes advantage of it and thrives, it does so unknowingly. But in the human case the potentialities of a chance technological discovery will be grasped, pondered, exploited, and made a springboard for further discovery.

So far, that is, as there remains any incentive to further discovery. I say this because most civilizations, notably those based on slavery, serfdom, or storable surpluses, have possessed a built-in ceiling or upper profitable limit to technological development. As

[3] *Pensées*, §1.
[4] See *I and Thou* (1923), tr. Ronald Gregor Smith (Edinburgh: T. and T. Clark, 1966).

Ernest Gellner has pointed out, where labour is plentiful and coerced, or its product easily confiscated, there is no incentive to economize on it.[5] So, a few basic tools apart, why bother with machinery, except for fun, ornament, or pure scientific curiosity (i.e. for the purposes of leisure)? I am thinking of such things as Yeats's mechanical songbird in his Byzantium poems, or the Greek inventor Hero's aeolipile, a hollow rotating vessel driven garden sprinkler—fashion by steam issuing from nozzles mounted on its circumference. Hero's was the first device to embody the principle of jet propulsion, which till then had been the property only of certain molluscs, but whose belated application has subsequently transformed the world. Yet jet propulsion would have been of no practical use or interest to the Greeks, who thought even foreign languages a waste of time.[6] And it is not obvious that jet propulsion has been an unmixed blessing to their descendants, when you consider the annual invasion of the Aegean by British lager louts.

The second key feature of technology is that it treats the world (which, as I have said, ideally or originally signifies only the natural world) as a means, and not as an end. Technology is the animating impulse of *homo faber*, productive, tool-using man. It not only works upon nature with instruments, it treats nature itself as instrumental to human purpose. And this of course is another respect in which it differs from science. The aim of science is theoretical understanding, irrespective of practical utility. The aim of technology is practical utility, irrespective of theoretical understanding. Technology cultivates theoretical understanding only so far as theoretical understanding contributes to present or future production (I am assuming here that production and utility are equivalent, as for the most part they will be in a demand economy). Whenever production can be improved without it, technology can perfectly well dispense with theory. Indeed, its intrinsically labour-saving tendency more or less guarantees that it will do so. Why bother with theory if you don't need to?

In identifying production or practical utility as technology's ultimate rationale we come back to the question of value. Value is of two kinds. Either a thing is valuable in itself, or it is valuable

---

[5] Ernest Gellner, *Plough, Sword and Book: The Structure of Human History* (London: Collins and Harvill, 1988).

[6] A first-generation disciple of Freud's, Hanns Sachs, claimed that despite their pre-eminence in science, mathematics and philosophy the Greeks were too inward-looking or 'narcissistic' to interest themselves in technology. I doubt, however, whether we need a fancier explanation than Gellner's.

because it enables us to obtain or achieve something else which is intrinsically valuable. In other words, a thing is valued either as an end or as a means. The value of the means derives from the value of the end, and is, moreover, secondarily determined by how efficiently it serves it. It is conceivable that some means, when absolutely indispensable to the achievement of a given end, may attract to themselves some or all of that end's intrinsic value and thus come to share its end-like character. (It seems possible, for example, that for Mill liberty was some such hybrid, half-means and half-end.)[7]

Value, therefore, may be either intrinsic or derived (instrumental). Certain objects will possess both kinds of value.[8] For example, basic literacy and numeracy are valuable almost entirely as means. Where the end is wrong, as in the case of libellous publication or crooked accountancy, it is accordingly possible to speak of such accomplishments as being subject to abuse. An end in itself, however, cannot be abused, because it cannot be used either. (I must ignore the exceedingly difficult question as to whether ends in themselves must necessarily be 'good'.)[9] Full intellectual culture, the Aristotelian contemplative life, is valuable almost entirely as an end (though Arnold, like Newman, believed that it also improved our manners). But most of the intermediate educational stages are at once necessary steps to higher levels, and plateaux of achievement which, though defective in comparison with the higher, are also still worth reaching for their own sake.

The secret of education, one might add *à propos*, is to invest as many levels of it as possible with this aura of intrinsic worth, particularly the earlier ones. Where this is not feasible, the student must simply be impressed as dramatically as possible with the worth of the end towards which his currently fatiguing labours tend. A child bored with scales and arpeggios will persist with

[7] The true ends of liberty, he tells us, are truth, autonomy, self-realization and the rest, but since none is attainable without it, liberty itself becomes almost an end (if indeed it is not one already).

[8] Though perhaps only when seen in different contexts. Here and now, it may be, in any single perspective, a thing is either a means or an end, but never both at once. We may be considering a duck/rabbit phenomenon.

[9] I have tried implicitly to answer it in 'Must new worlds also be good?', *Inquiry*, Vol. 38 (June 1995; (forthcoming at the moment of writing). The answer, such as it is, is largely in the negative, at least in the case of ends in themselves which are no more than subjectively perceived as such. (For example, torture, for a sadist, might very well be considered an end in itself. But it could hardly be so for his victim, unless by a happy coincidence his victim happened to be a masochist.)

181

them once he is accustomed to hear proper performances of real
music, and understands the connection between his drudgery and
the performer's art. In fact, the sooner he is given some real music
of his own to perform, no matter how simple, the more obvious
the connection will be.[10]

Roughly speaking, technology means the devising of 'solutions'
to independently specified 'problems', the search for optimal
means to given external ends. It is not and could not be competent
in an even more essential human activity, the choice or scrutiny of
those ends. Nevertheless, as a skill, it contains its own intrinsic
satisfactions, and so far (but only so far) may be said to constitute a
value in itself.

For an exaggerated example, consider a crossword puzzle. Here
the point is not the actual solution (as it is in technology), but the
intellectual challenge and effort involved in finding it. In a crossword puzzle (as opposed, say, to a piece of police detective work)
the apparent means has become the real end. A detective will take
any route to his goal, and the shorter and quicker it is the better.
He will be grateful for undercover tips, and overjoyed by an unso-
licited, fully corroborated confession. By contrast, though the
crossword normally takes me a whole hour, and I know that my
friend Snooks can usually do it in ten minutes flat, I should nevertheless think it peculiarly pointless to ring him in order to get the
answers. If the solution were the real object, however, Snooks
would be my best and most economical route to it.

Technology, like any other skill, may exert this kind of aesthetic
or athletic fascination. (It has often dazzled intellectuals incapable
of tying their own shoelaces.) To that extent it will come to resem-
ble an end, or (as we might say) a value, in itself.[11] And, when this

[10] I would like to say how well Bartók's *Mikrokosmos* fits this particular
bill, but—alas!—most children find it as dry as scales, just as they preferred
Barbie dolls, fluorescent plastic dumper trucks and (horror of horrors) My
Little Pony to those Arts-and-Craftsy varnished wooden efforts (by Galt
Toys, Paul and Marjorie Abbatt and the like) with which their enlightened
parents once endeavoured to stimulate their imaginations. (A matchless
chronicler of these and similar middle-class agonies is the *Guardian* cartoonist Posy Simmonds, as was another cartoonist, the late Mark Boxer,
with his 'Life and Times in NW3' in the now defunct *Listener*.)

[11] Ends in themselves, it should be said, and perhaps paradoxically, are
generally perceived as being those goods, activities or relationships in
which human beings achieve maximum all-round fulfilment. (See also
note 9 above.) If a thing is an end in itself, it is supposed that it must
belong, not to a hierarchy, but rather to a spectrum, of ends. Every
unqualified end in itself stands at the apex of a hierarchy of intermediate
ends, which in relation to itself are means.

occurs, it will not always or necessarily be a bad thing. Indeed, the absence of any such 'spiritual' satisfaction or challenge in one's work is very largely what provoked nineteenth-century thinkers from Carlyle and Coleridge to Morris and Marx to criticize the new industrial capitalist order.

Nevertheless, the primary value of technology is instrumental, and is determined by its ultimate end. So far, but only so far, as that end is to free us from drudgery, thus liberating us, if we wish, for more creative and less alienating pursuits, technology is almost wholly to be welcomed. Who would not rather own and use a washing machine and tumble drier than spend, or see his wife spend, a couple of days a week on the family's laundry, as in Edwardian times (and even in my own childhood)? Of course technological advance, like most goods, does not come without a price tag, even if we leave out of account the temporary unemployment of labour made obsolete by new technology, as also the fact that some of our fellow-citizens appear to fill their increased leisure with activities (or should one say passivities?) of stupefying vacuity. I am thinking rather of such things as waste disposal, the depletion of energy resources, or indeed (if that problem were to be solved, say by nuclear power) the thermal build-up resulting from increased energy use.

Technology has other costs too, as spoilsports such as the late Fred Hirsch (*The Social Limits to Growth*) seem almost indecently eager to point out. Until nearly every family had one, the motor car, like the railway, represented an enormous extension of human possibilities. But the social costs and practical inconveniences of present-day motoring have nothing to do with its crude, or so to speak abstract, technological benefit. They have to do purely with its being what Hirsch tendentiously calls a 'positional good'. In thinly populated areas, however, it is not a positional, but a simple, unqualified good. There, running costs apart, the disadvantages of car ownership are negligible, whereas those of non-ownership are enormous.

On balance, I should say, technology's most outspoken critics are either in error, or insincere, or possessed (just as its most fanatical enthusiasts once were) of a hidden, *dirigiste* agenda. Certainly they are rarely to be seen denying themselves any of its genuine benefits, especially in the medical sphere. However, there is a really serious criticism to be made, not of technology *per se* (for how can a means be criticized, except in point of its adequacy to its end?), but of its perversions. Strictly speaking, technology as such can hardly be perverted. Only its end can, and that, being external to it, must stand or fall by independent, non-technological criteria.

We do not fault a hammer, say, for being improperly used as a murder weapon. We lay the blame where it belongs, with the murderer (or perhaps, as D. H. Lawrence would say, with the murderee). Nevertheless, to pursue a line of thought begun earlier, there exists what we might call an uncritical spirit or culture of technology. It is not part of technology itself, but consists rather in a particular climate of opinion concerning it and its place in human society. I refer to what is now widely called 'technocracy'.

This word was coined in 1919 by the engineer W. H. Smyth, an admirer of Veblen, to denote the so-called, and by him eagerly awaited, 'rule by technicians'. Nowadays it means something more like rule by believers in technology, especially those who believe in the extension of technological methods and assumptions to social and political matters and the conduct thereof. This is the real perversion of technology: the application of it to human rather than material objects, and the consequent necessity of treating humanity as an unqualified 'It' (against which, of course, Kant's Second Categorical Imperative explicitly warns us). A further, even more sinister perversion must inhere in the substantive, virtually material transformation of the human object, as found in (say) chemical technologies of mind and mood control, certain kinds of genetic manipulation, etc.[12] All this was foreseen by Aldous Huxley in *Brave New World*,[13] and amounts to what C. S. Lewis called 'the abolition of man', meaning of man as we know him (that is, as a being both free and sinful, which is to say moral).

The technocratic ideal was not in fact new. In one sense it was a century old: one thinks of the early utopian positivists and social engineers such as Comte, Saint-Simon, and Fourier, many of whom actually were real engineers by profession. Such people spoke quite blithely of what they imagined to be the coming transformation of human nature, as indeed did Marx, who was by no means as remote from them as he liked to think. A nature which

[12] A distinction should be made between the use of chemicals (a) to restore the normal balance of mind in psychiatric patients and (b) to upset the normal balance (as formerly in the psychiatric 'treatment' of Soviet dissidents). The same distinction holds between genetic engineering used to relieve hereditary disease in individual cases, and the same employed wholesale, say to produce a race of obedient Huxleyan 'epsilons'.

[13] A complex, scrupulous and rather pessimistic attempt to devise a naturalistic ethic for the technological age in the light of its increasing Huxleyan possibilities and ecological threats is Hans Jonas, *The Imperative of Responsibility* (Chicago and London: University of Chicago Press, 1984).

can be wholly transformed is tantamount to pure artifice, that is, to no nature at all. Everybody agrees that natural rights and the like are contentious, but it seems clear that if they exist, they can attach only to something which belongs to nature (since it is nature which confers them), and therefore also *has* a nature. In a good many utopias humanity will be so radically transformed as to stand of no need of rights, and not merely because all possible occasions of conflict will have been removed, but also because, since none will desire freedom, there will be no freedom to protect.

In another, only slightly extended sense, technocracy harks back to the rule of experts recommended by Plato. His Guardian class is not literally made up of technicians or engineers. But their right to rule is supposed to derive from their rationality, of which, for Plato, the ideal type is, if not exactly instrumental, at any rate mathematical. For Plato moral and political knowledge, the knowledge of intrinsic values, is barely distinguishable from mathematical truth. Accordingly mathematics forms the chief ingredient of the Guardians' education.

The liberal relativist objection to Plato, to the effect that moral and political values are not and cannot be firm, objective or absolute, in the way that mathematical truths are supposed to be, is sound enough as far as it goes. It might, however, seem to be true that the only conceivable use for human imperfection, not least as evidenced in the imprecision or just inherent fuzziness of our goals, conceptions and proceedings generally, must lie in its ministering to, or being instrumental in the realization of, that which is perfect (assuming we know what it is). But the weightiest objection to Plato is, I think, this: that even if moral and political values were as mathematically unequivocal as he supposes (and certainly, like aesthetic values, they pretend to objectivity of a kind), a ruler has no business trying to impose them on his subjects, or setting out to realize them in some imaginary perfect social end-state. It is not that he lacks the right (which, being a deliberately selected, literal aristocracy of the 'best', the Guardians must possess pretty well by definition), but simply that he will fail, and also wreak an immense amount of havoc before he realizes he has done so.

To be sure, socio-political values are ends of a sort. They are not means, certainly, or if they are, they will hardly be effective (in fostering social cohesion, say) if they are frankly treated as such. For to see them in an instrumental light is to reduce them to mere prudential maxims or hypothetical imperatives, and thus to strip them of their moral force. But to call them ends is not to say that they can be finally realized in some programmatic, quasi-technological 'solution'. Perhaps the term 'end' (or goal, or *telos*) is after all a misleading synonym for everything that possesses (or should one say constitutes?) an intrinsic value. For it is hard to see moral values in the light of a substantive purpose to be accomplished, without sensing a category mistake somewhere. Moral values, normally, are not things we actively pursue,[14] but rather conditions to which, in our pursuit of other goods, we are obliged to subscribe.[15] They are not so much a goal as a guiding principle.

An individual may set out to amass a fortune, be a concert pianist, or achieve religious salvation, without falling into obvious absurdity. Even a society may reasonably pursue certain modest collective goals (winning a war, say). But it is not clear that either an individual or a society can actively pursue happiness, or the good, or justice. Those things are certainly ends, in the sense of not being means, of being intrinsic values. But it is doubtful whether, without a certain sentimentality or self-indulgence, they could be self-consciously entertained as goals. It is more likely that they are the outcome, or reward, of pursuing something else (duty or others' interests, for example), or of governing one's conduct by certain general moral considerations or side-constraints. Justice, of course, belongs to this latter category (only utopians *aim at* justice, in the sense of a comprehensive end-state, and in so doing are constrained to commit much injustice along the way); the good, however, is rather too general to serve as a moral benchmark; while happiness or *eudaimonia*, though it may be the end of virtue (so long, that is, as virtue is pursued as an end), cannot in itself generate any specific constraints. Though it may result from vice, unhappiness is not itself a vice, nor to be avoided for that reason.

Morality, then, though both intrinsically valuable and conducive to human happiness (which is why moral behaviour is rational) is not properly to be understood in quasi-technological terms, as a goal. (I will not need to enlarge on the observation that utilitarianism, however persuasive it may be in furnishing the underlying principle of certain usually military or administrative decisions, is the moral technology *par excellence*.) And there are further reasons why technology is an incomplete or inappropriate model for conduct generally.

[14] In *The Wild Duck* Ibsen depicts the kind of fanatical idealist (Gregers Werle) who believes that they are, and the disastrous consequences for others of acting upon this belief.

[15] Oakeshott's *On Human Conduct* (Oxford: Clarendon Press, 1975) offers a reading of morality in such 'adverbial' terms (see pp. 60–81. esp. p 66). There is conceivably an affinity with Nozick's conception of morality as a set of 'side-constraints' (see main text, below).

One has already been indicated. It will be helpful here to refer to the greatest of all political quasi-technologies, Marxism. Under communism, said Engels, the government of men will be replaced by the administration of things. The phrase was not happy, since what communism actually did was turn its subjects into things. That, however, is just what technology demands, and why, so far as its materials really are material (sc. passive) it succeeds within its own narrow sphere of competence. But that is also why (to say nothing of the tens of millions of corpses) its political extensions have always been a resounding failure. No political technology can take account of people's propensity to prefer their own goals, projects and values to those which unwise rulers have defiantly elected to impose on them. And it can only discount this propensity by extending sufficient coercive power to suppress its manifestations, an enterprise which genuine, material technology, in the shape of weaponry, electronic surveillance, and so on, has proved only too ready to assist (Wise rulers, of course, will as far as possible support their subjects' goals, projects and values. Political consent, as opposed to mere sufferance, is always reciprocal. The good ruler 'consents to' his subjects' values and preferences, and that is why they consent to his government. They see it as representative of themselves, whether or not any electoral mechanism exists to make it so formally.)

To say that technology, which is after all in one sense purely impersonal, a mere tool, is nevertheless 'only too ready' to assist in the work of repression recalls us once more to the fact that it is actually a human activity, an expression of human will, which, like all others, is subject to moral permissions and constraints. It is precisely because technological processes and procedures are neutral—that is to say, employable in principle in any cause, good or bad—that they, or more accurately the ends they subserve, require moral scrutiny. That is simply to say that no matter how efficient, elegant, productive or functionally beautiful a technology may be, it is no more self-justifying than any other means, and is always subject to Solon's demand that we should 'regard the end', or at least not violate superior ends (call them rights if you must) in pursuing it.

And the same goes for bureaucracy, which is simply and inescapably the technology of government in advanced societies. There is a radical difference between the so-called 'desk killer' Adolf Eichmann (to whom Hannah Arendt devoted an entire book) and a now endangered species, the traditional British mandarin. It is not just the difference between technological impersonality and civilized impartiality, nor that between blind, obsessive obedience and responsible, educated loyalty. Nor is it simply the difference between the orders each was or is likely to receive, from a Führer on the one hand or the Crown in Parliament on the other, the difference, one might say, between arbitrary power and properly constituted, legally limited authority. It is something to do with the perception that, though specialization and hierarchy (the division of labour, in other words) are necessary for the smooth functioning of government, they are not by themselves sufficient. For civilized government, more is required, viz. culture or liberal education, which is a thing much wider than government itself.

We are speaking now of a knowledge beyond that of means, namely a knowledge of ends. If we ask in what it ought to consist, we could do worse than look at what our Victorian forebears, the founders of the modern civil service, prescribed for their rulers and administrators. Classics, mathematics and sport; perhaps we would not nowadays seek to duplicate the recipe exactly. What is disconcerting, however, is the fact that, given a few adjustments, these are substantially what Plato put on the Guardians' curriculum. Maybe Plato was less totalitarian than we have been led to believe. Or perhaps his mistake was to imagine that the Guardians, though selected and educated according to his prescription, would necessarily wish to rule over the Republic as he had originally conceived of it. If truly wise and humane, they might have seen the sense in limiting their own discretionary power, establishing a rule of law and governing in accordance with it. They would doubtless also have abolished the special conditions imposed upon their own lives, sexual communism and the ban on trade, not only as onerous, but also as too obviously separating them from their subjects. In short, they might have turned the original Republic into something like a civilized liberal state, and been content to live with its inevitable imperfections.

Question Time*

RENFORD BAMBROUGH

It makes straightforward sense to ask a person 'Why did you decide to become a solicitor?' (or a carpenter, shopkeeper, teacher or professional musician). There is no mystery about the question 'Why did you decide to become a British citizen?' (or a member of the Liberal Party, the Roman Catholic Church or the Aristotelian Society). It may be difficult to answer any of these questions. We may not remember, or may be unable to articulate, what first led us to seek ordination, join the Army or stand for Parliament; but the questions themselves are clear, and they ask for fair comment on matters of public or private interest. There are other questions that sound like these but raise difficulties of another order.

'What made you decide to become a woman?' This question can be intelligibly asked only of a woman. It cannot be asked of any woman, or of many women. For it can be asked only of a woman who has decided to become a woman. But has there ever been a woman who has become a woman by her own decision? What the newspapers call a sex-change is understood by the new woman as the revelation and acknowledgement and confirmation of a womanhood that was there from birth, unrecognized or disguised and disfigured by shame or shock.

'What made you decide to become an Englishman?' This question could be sensibly addressed only to a man who had decided to become an Englishman, and there is no such animal. It is by birth that an Englishman is an Englishman, and by birth that a woman is a woman.

It makes sense to ask of some British citizens why they decided to become British citizens. It makes sense to ask of some British citizens why they have decided to remain British citizens. It makes sense to speak of person's having been a British citizen and having ceased to be a British citizen, or vice versa, and accordingly it makes sense to ask of some people why they have decided to be or not to be British citizens. But the typical British citizen has never decided to be or to remain a British citizen. The question has never arisen.

Being a British citizen is a birthright that I can lose or renounce. Being an Englishman or a woman is a birthright or birth-burden

*Philosophy in Britain Today, S. G. Shankar (ed.) (London & Sydney: Croom Helm, 1986).

that its possessor or bearer cannot lose or renounce. You cannot by taking thought become or cease to be English or female any more than you can by taking thought add a cubit to your stature.

The short answer to the question 'What made you decide to become a philosopher?' might be the question 'What made *you* decide to become a human being?'

Being a philosopher is a birthright too, but one that it is possible to renounce or lose. It is not however a right or burden that it is possible to choose to acquire. It belongs to all human beings by birth, though many afterwards lose or renounce it. Nobody can acquire it except by birth.

This is not to deny, but to put in their place, some familiar facts that I might seem to be forgetting. There is much that is connected with philosophy that may be sought and found, acquired and lost. I may seek or find recognition as a philosophical author or lecturer. I may pursue the profession of teaching or writing philosophy. Ryle boasted and confessed:

> I myself am not ashamed but, when I happen to think of it, slightly complacent to have made, as a philosophy don, my own living—and quite a good living—for half a century; and I am not ashamed but thoroughly proud, when I think of them, to have achieved, as a philosopher, certain offices and distinctions and to have exercised a bit of influence, authority and even modest powers.[1]

Doing these things is one mode of life that may be called that of being a philosopher. But these things are peripheral and inessential. None of them was done by Socrates. Philosophy is a vocation before it is a trade or profession, whether in the life of an individual thinker or in the history of human thought. Plato in the *Meno* and Wordsworth in the 'Ode on the Intimations of Immortality from Recollections of Early Childhood' remember and celebrate this truth about the human condition. Ryle does not sufficiently remember it because he is frying other fish. Societies for the promotion of philosophy for children do remember it but misconstrue its significance. Hare in 'Philosophical discoveries'[2] and Wittgenstein in most of his work propound and enact it, and do so without falling into sentimentality or any of the other hazards that surround it.

Wordsworth puts it romantically, Plato classically, and Hare more prosaically, but what they all remind us of is the value of keeping open our lines of communication with our earliest learn-

[1] Gilbert Ryle, 'Fifty years of Philosophy', *Philosophy* (1976).
[2] *Mind* (1960).

ing. The need for this preservation of links with childhood is not confined to philosophy. By birthright we are all not only thinkers but also singers and dancers, poets and painters, teachers and story-tellers. This means that the professional singer or painter, poet or teacher, dancer or story-teller, is a professional in a different way from the solicitor or doctor, physicist or statistician. The same is true of the professional runner or jumper, rider or speaker, gardener or interior decorator, soldier, nurse or politician.

Like the runner of the writer or the ruler the thinker may become a professional but can never become an *expert*. Even the geniuses among thinkers or writers—Shakespeare and Tolstoy, Plato and Wittgenstein—are doing to a higher power something that we all do and need to do for ourselves. This affects the light in which the thinker is regarded and the role to which he is called. The mathematician and the physicist are allowed to be technical and abstruse. The philosopher and the poet, the sculptor and the novelist, are expected to remain in touch with their roots, which are also our roots. It is they and not the statistician and the bio-chemist who are accused of obscurantism and mystification when they fail to communicate with the ordinary understanding of the ordinary person.

When I heard Wittgenstein at the Cambridge Moral Sciences Club in 1946 and 1947 I had already formed, even if I could not so fully articulate, some of the convictions about philosophy and its role that I am setting out in this essay. Wittgenstein's sayings, and later his writings and the sayings and writings of John Wisdom, strengthened the convictions and assisted their articulation. In particular, Wittgenstein's reminder of the importance of reminders, his insistence on the need to 'go back to the teaching', reinforced my confidence that philosophical thoughts were communicable by non-technical means, that its points of reference were landmarks familiar to human beings in general, even if I had not heard it said before that philosophy simply rearranges what we have always known, that it leaves everything as it is, that it offers no theories or explanations, brings us no news.

In those same years, as an undergraduate reading for the Classical Tripos, I was reading Plato and noticing not only the formal doctrine of *anamnesis* and the declaration of the *Phaedrus* that each of us has seen the *onta*, the eternal Forms, but also the Socratic aims and methods that were sometimes disguised by the formal and metaphysical clothes in which they were dressed by Plato. Socrates talks to one person at a time, and says that he is concerned only with what is in the mind and heart of that one person. His *elenchus* is the examination of the hitherto unexamined life and mind of Polus or Callicles, Glaucon or Adeimantos, Protagoras or Meno. The tools and materials of the examination are drawn from that mind itself, and the end in view is the modification by self-examination and self-knowledge of the same mind. The destination of a thought is as important as its origin, and Socrates sees the origin and the destination as the same.

Socrates and Wittgenstein both provide precedents for my answer to the question about my choice of topics to work on. Like them I do not see philosophy as broken up like history or biology into departmental specialisms. Even if the questions of philosophy arranged themselves into such patterns of genus and species, it would still not be a matter of choice to be interested in these topics rather than those. The philosopher is inescapably a philosopher by being inescapably, involuntarily, puzzled or perplexed. Experience suggests and theory confirms that the questions by which he is puzzled are interconnected in ways that make the task bewilderingly difficult but also make it possible to tackle it with some hope of progress. Each question raises new questions, but the other side of that coin is that when we are concerned with any one question we are likely to be able to orient ourselves for the exploration of it by remembering or being reminded of what we already understand of a matter that we can see to have a bearing on our present perplexity.

Both the puzzles and the landmarks are familiar to us from childhood. Wordsworth spoke of the Immortality Ode about those obstinate questionings 'of sense and outward things' that make the infant child into a philosopher. The questioning is at first of what is *not* seen: 'Is our house still there when I am at school?' Later what is seen is also questioned: it could *look* like that even if it were really not at all like that.

The unity of the network of questions is liable to be disguised from us because the network, though unified, is also highly ramified. The maze looks different at different points of entry, but from any point of entry or any point within the maze it is possible to travel to any other. The philosophical maze does not have edges or boundaries like the Hampton Court maze, and it has no unique solution, but its structure is otherwise similar enough to warrant Milton's use of the image in Book II of *Paradise Lost*, where devils in hell are 'in wandering mazes lost' when they debate 'fix'd fate, free will, foreknowledge absolute'.

I was in this maze from early childhood, as most children are. Besides the questioning of sense and outward things I remember being puzzled about past and future, freedom of action, causality and foreknowledge. All these, like all other interesting questions of philosophy, are central to the 'subject'. They are not chosen as topics to work on but presented as difficulties to struggle with.

This is not to deny that the presentation of the difficulties, and to some extent the form of the difficulties presented, may be influenced by events and circumstances of individual lives and particular centuries and generations. This may be the right time and place to supply, as requested, 'some autobiographical material'.

I grew up from the age of 13 to the age of 19 during the Second World War. Until 1939 my family lived in a Durham mining village, and by the end of the war I was working as a coal miner under the 'Bevin Boy' scheme. At such a time and in such an environment I soon knew the difference between being intelligent and being educated. I was surrounded by people of all ages who deserved but had not received any advanced education. Soon after the war admission to Sixth Forms and colleges and universities became more systematically 'meritocratic', but when I was an undergraduate at Cambridge from 1945 onwards there were still many who were there mainly because they could afford it, and who did not try to compete academically with the poor scholars of their generation.

Some of the miners I worked with were keen to discuss politics and philosophy. Those of my own age knew that I knew things that they had had no opportunity to learn, but they also recognized that on many of the questions we talked about there was no specialized knowledge to be had. On the rights and wrongs of socialism, communism, conservatism and liberalism, war aims and post-war planning, we were better matched than when I spoke of nineteenth-century history or Plato's *Republic* or the origin and structure of the United States of America or of the solar system. (These are all typical of the things that my fellow-workers were interested in and asked me about.)

I had already learned from school debating that people of comparable intelligence and education could disagree about important questions of ethics and politics and religion. It was through debating too that I first came to hear the word 'philosophy'. I competed in a Prize Debate on the proposition that 'The Best Things in Life are Free', and the adjudicator described the argument of my speech as philosophical. He was E. I. Johnston, the senior Classics master, and so I had opportunities of hearing much more from him, and of learning how much of what I was interested in fell under that same rubric. Later he took me through Book I of the *Republic* as a Higher Certificate set book, and in that and many other ways extended a debt which I partly discharged by dedicating a book to him. Though I stayed with the Classics not only as an undergraduate but for nearly 20 years afterwards, when I left his hands I was already convinced and confident that my main work would lie in philosophy.

At Cambridge the debating continued. Free will and determinism over coffee in the small hours. Politics of the Union Society and in the political clubs. And now in meetings and debates and lectures I could hear and assess speakers, natives or visitors, whose life work had been to think on these things: Bertrand Russell, Hugh Gaitskell, Quintin Hogg, Isaiah Berlin, John Strachey, C. D. Broad, Herbert Butterfield, Michael Oakeshott, Karl Popper, A. J. Ayer, Gilbert Ryle, John Wisdom, Ludwig Wittgenstein.

It was not rare in those days, at least in Cambridge, to practise philosophy without reading or teaching the subject officially. Most of those who crowded into the Moral Sciences Club when Wittgenstein was chairman were students or teachers of other subjects, and all but one of the professors and lecturers of the Faculty of Moral Science had come into philosophy from other fields.

I have lived on into a generation in which the vast majority of teachers and writers of philosophy are conscious and proud of being trained professionals, and who understand the idea of a profession by analogy with that of the physicist rather than that of the writer or critic. (My colleagues were almost unanimously delighted when an empirical psychologist told them that Cambridge philosophy students, unlike recruits to psychology from other arts subjects, 'think like scientists'.)

An Oxford philosopher, asked recently by a visiting lecturer in English what the young philosophers are interested in nowadays, said: 'They are interested in rigour.' But they are not. They are interested in technicality, which is quite a different thing. Philosophy is thinking which is rigorous but not technical:

> The language of philosophy is therefore, as every careful reader of the great philosophers already knows, a literary language and not a technical. Wherever a philosopher uses a term requiring formal definition, as distinct from the kind of exposition described in the fourth chapter, the intrusion of a non-literary element into his language corresponds with the intrusion of a non-philosophical element into his thought: a fragment of science, a piece of inchoate philosophizing, or a philosophical error; three things not, in such a case, easily to be distinguished.[3]

Philosophers should teach and preach that it is possible to be accurate and precise and exact without being either professional or

[3] R. G. Collingwood, *An Essay on Philosophical Method*, pp. 206f.

technical. Instead they are liable to ape the external forms of mathematicians and natural scientists, or at least linguists and technical historians, in order to avoid appearing 'literary' and hence unprofessional. Their idea of philosophy is of *normal* philosophy, like T. S. Kuhn's 'normal science'. Even if they are interested in Wittgenstein's paradigm-shifts they do not see that the whole enquiry of philosophy is a matter of paradigm-shifts—of seeing one thing as a case of another—for example, seeing a river as a fast-changing mountain and a mountain as a slow-moving river. Accordingly, the paid and trained philosophers who write about Wittgenstein turn him into one of themselves, a member of a community of ideas, a practitioner of techniques and procedures of argument or of scholarship that do not belong to the general conversation of mankind from which philosophy arises and to which it must return, as Socrates knew, from its most or least adventurous excursion.

These roots in the common understanding help to account for some familiar but sometimes forgotten facts. Philosophers take easily to questions of policy and administration. There are many philosophy graduates in the Civil Service and local government, hospital administration and similar services. Teachers of philosophy often turn to university or college administration, or combine it with their philosophical work. Members of the 'profession' are in demand as chairmen or members of government committees and commissions. What makes this natural and appropriate is that questions of policy and politics and practical ethics share their structure with those of philosophy: they are complex but *informally* complex, like the dilemmas and difficulties of ordinary life and not like problems that yield to well-disciplined formal thought or well-directed observation or experiment. Lawyers typically share an understanding of these points, since their own training is largely a training in how to understand and articulate tangles of ideas and concepts, though their institutionalized need for a yes-or-no answer for practical application sometimes makes them more rigid than rigorous in the sense of rigour that I am concerned to defend.

Graduates in philosophy are in demand by employers in the computing industry, not because they think like scientists but because they understand the need and scope for formal thinking like that of the scientist and the mathematician without forgetting the possibility and value of a logical discipline that is not so easily if at all reducible to rule. (I have found when teaching philosophy students in pairs or groups, in the Cambridge system where it is common for students to change to philosophy from other fields, that the ex-mathematician can show the ex-historian or classical or literary scholar the value of formal thinking, while learning from them in turn that not all worthwhile thinking is formal.) Philosophical training and experience are also found among graduates employed in journalism, authorship, publishing, broadcasting and television and other 'media' careers, where again there is a need for flexibility as well as force.

I have engaged in a number of the activities covered by these last remarks. Most of my broadcasting work has been philosophical, but it has usually been at points of contact with other fields and with ordinary life and the common understanding. A typical example is 'Nature and Human Nature', a series of duologues in which I interviewed eight specialists in the human and behavioural sciences—anthropology, sociology, psychology, medicine and linguistics. The participants were not much practised in philosophy, but this did not unfit them for the exercise of articulating—with a freshness that came from doing it for the first time, and in response to questions which were simple but sometimes unexpected—the assumptions on which they thought and acted in the pursuit of their professional objectives. My own role was very similar to that of a philosophy supervisor (tutor) in the Cambridge teaching system, which is best fulfilled by asking short and simple questions that may call for long and elaborately qualified answers.

That is also the Socratic role; the habit of noticing contradictions or confusions between what is said by the same person on different occasions or even on the same occasion. This is the mode of examination by which Socrates aspired to turn an unexamined life into an examined life. The same process proved appropriate to two other areas of my supposedly non-philosophical work, and especially at a juncture when the two areas were for a year or two closely similar in their needs and problems. From 1964 to 1979 I was Dean of my college. From 1964 to 1970 I was a member of the Council of the Senate, which the *Cambridge Evening News*, comically but not altogether misleadingly, always calls 'the University's "inner cabinet"'. In both capacities I saw a good deal of the 'student troubles' of the late 1960s and early 1970s, which were a question time for universities.

Most of the rebellious students were inarticulate henchpersons for a small group of leaders exercising a leadership that did not officially exist. The leaders were mainly acute and intelligent questioners, with whom it was interesting for a philosopher to discuss topics of the day and the hour. They themselves would have liked it to be topics of the minute: a rolling programme of perpetual revolution was one of the watchwords of the time. Yet even the leaders were hidebound by their little revolutionary handbooks, and acknowledged that they were disconcerted to find a college Dean, who ought by the book to have been confined to the role of 'bureaucratic formalist', asking them questions more far-reaching than those concerned with the latest demonstration or breach of rules. The College radicals solemnly considered at one of their meetings a motion of protest to the Master and the College Council that it was unfair to them that the office of Dean should be occupied by a philosopher, a person more at home than they were in arguments on matters of principle. To my disappointment, one of their number was alert and prudent enough to suggest that they were in danger of making themselves ridiculous, and the motion was withdrawn.

The discussion of these high matters could be continued in college societies concerned with philosophy or history or literature or religion. I was also a member of a university committee of senior and junior members set up to review the disciplinary regulations of the university. In all these contexts it was noticeable how much practical importance can attach to the discussion of what look like abstract theoretical issues. On the senior as well as on the junior side there were many whose grasp of the issues was implicit or insecure, and who needed and welcomed the help of spokespersons with some taste and zest for the dialectic.

I have dwelt on these relatively trivial events because I believe they illustrate in their small scale the role of a philosopher in a community large or small. What I represented to the students was not only my own idosyncratic opinions but the unspoken sense of purpose of a college and university community. The challenge of the rebels was a demand for an explicit justification of every rule or practice that was ever questioned. When it transpired, as it inevitably did, that such a process of attempted justification came fairly abruptly to points at which it was hard to avoid the appearance of mere dogma, the rebels were encouraged and their opponents, junior as well as senior, bewildered by the experience, familiar to every victim of the Socratic gadfly, of knowing that what he is asked to agree to must be wrong, yet being unable to *say* what is wrong with it. If the structure of the issue can be more fully articulated and explained, the original challengers will be little if at all affected. They are not interested in finding the answer but in making capital out of the question. But the bystanders—who are usually in such cases the bulk of the population—will feel better armed against what they recognize to be a threat to more than their peace of mind.

Political and moral philosophy as applied to the life and work of the larger community are writ small but legibly in these parochial events. 'Society', like an individual institution or tradition of thought or teaching, is a going concern whose basis may not be capable of being rendered explicit in the way demanded by its critics. But its critics are themselves beholden to it for the sources of their criticism and of their constructive offerings if they have any to make.

What is wider and more fundamental still is that *all* reasoning is a going concern in the same sense, and subject to the same difficulties. We did not create our society by act of will or by adopting a set of conventions. We equally did not at any time adopt principles or practices of thought and speech. An act of criticism, theoretical or practical, is an act that belongs within a tradition of theory or practice. What Wittgenstein dangerously calls our 'form of life'—our whole human existence—is transmitted and inherited and modified, but it was not made by us and cannot be renounced by us.

This brings us back to my 'own work' in the only sense in which I am prepared to regard it as professional. In the first place it speaks of the *role* that belongs to a philosopher, one which invites and provokes a special attitude towards him on the part of other people. He is their critic but also their representative. He speaks against them when he exercises his talent for noticing contradictions, but he speaks for them when he invites them to resolve their contradictions by being more faithful to the coherences of the inherited human understanding. The double process prompts a double attitude, or an oscillation: the victims and beneficiaries may welcome his help or resent his interference; they may crown him with laurel or dose him with hemlock.

But there is another and more familiarly professional way in which these examples return us to my philosophical work, past, present and future. I have been concerned with Wittgenstein, and with concepts of scepticism and justification; concerned with Wittgenstein because I have been concerned with scepticism and justification. The question why I have *chosen* to be concerned with these questions is adequately dealt with—though for reasons given there, not straightforwardly answered—in my earlier remarks about choice and interest and choice of interest. But I can now answer another of the editor's questions: why do I regard the topics that I have been concerned with as important? My answer will also explain why I intend and expect to continue to be concerned with them, and this whole package can be wrapped in an account of some work in which I have long been engaged, and which has raised for me some of the questions now raised by the editor.

I am approaching the closing stages of a project launched by a paper given to the Moral Sciences Club in 1962 and published in *Philosophy* in 1964. The title, '*Principia Metaphysica*', was deliberately chosen to imply an ambitious objective. The paper was a prospectus and a manifesto for a wide-ranging view of philosophy and for the book in which it would eventually be more fully expressed.

'*Principia Metaphysica*', though designed to be a draft of the first chapter of the book, was itself in origin a sequel to 'Universals and Family Resemblances', published in the *Proceedings of the Aristotelian Society* in 1961. This earlier paper was nearly but not quite explicit about two of the purposes of the whole programme: to express the main results of Wittgenstein's thinking more clearly by setting them out more systematically than he or his disciples had thought reasonable or even possible; and to demonstrate the continuity between his work and that of his predecessors.

It is again becoming fashionable to question the unity and continuity of philosophy from its Greek origins to the present day. In particular, it is questioned whether *epistemology* is a permanent element in the philosopher's preoccupation. Some suggest that it originated with Descartes, or died with Wittgenstein or Heidegger or Frege or the pragmatists. In its place, we are sometimes told, we must install or have installed something else, called ontology or philosophy of language or 'edifying' discourse—there is greater unanimity about the disease than about the remedy. The diagnosis and the prescriptions seem to me to be misconceived. All these physicians have allowed themselves to confuse differences of fashionable idiom with differences of philosophical substance. The use of a linguistic or logical or epistemological or psychological or ontological clothing for philosophical reflections commonly indicates a distinctive view of the nature of philosophy, but it ordinarily disguises only thinly a concern with a set of relatively permanent philosophical preoccupations. The most central are those which, when clothed in the epistemological idiom against which we are now so sternly warned, are summed up in the question 'What if anything do we know, and how do we know it?' It may be expressed in a number of other familiar idioms: What is real? what exists? What are the basic features of the world or the universe? How is language possible? What is thought or understanding?

Wittgenstein was as suspicious of the questions, in any of their traditional forms, as he was of the theories or theses in which philosophers had undertaken to supply the answers. But his repudiation of traditional forms of expression, which caused himself, his critics and his disciples to exaggerate the novelty of his work, was itself another of the changes of idiom on which philosophers base their periodical cries of revolution. There was a polemical and therapeutic point to Wittgenstein's announcement that he was engaged in 'an activity that is one of the heirs of the subject that used to be called philosophy', but every such paradox calls for redress after it has redirected our attention. Wittgenstein was a philosopher, and hence an epistemologist. This becomes clearest in his latest work, *On Certainty*, where he explicitly discusses the concepts of *knowledge* and belief and doubt and certainty, and the unmistakably and unashamedly epistemological work of Moore, and yet is largely faithful to the methods and results of the *Philosophical Investigations*. Like Moore, he is concerned with the common understanding, and with the demonstration that it is an *understanding*, and that it is *common*—shared by all human beings, and by all beings who could intelligibly be conceived as rational.

This recognition of the unity of humankind and hence of the human understanding again links Wittgenstein with Socrates and Plato, and also with Aristotle, for whom we are *logika zoa* (talking and thinking animals) as well as *politika zoa* (animals adapted and disposed to live in communities). Plato's picture of these unities is painted in the *Phaedrus* myth, where the gods impose an important constraint on the transmigration of souls between animal and human bodies and lives. No soul may embark upon a human life unless it is one of the privileged and enlightened souls who have seen the Forms. When this is translated into plainer prose by Wittgenstein it amounts to saying that all human beings share a background of understanding which is the source of all their questionings and all their answerings; a stock upon which we can all draw in our internal or external conflicts and perplexities. The Socratic examination is designed to determine which of the conflicting *logoi* best coheres with the fundamental knowledge that is shared by all of us. An important corollary is that to be in error is also to be confused, since it is to be in conflict with oneself as well as with the truth. The truth is in us, and when we deny it we are literally contradicting ourselves.

In the re-presentation of Wittgenstein's results that I am offering it first seemed to me that I should need to be not only more systematic than he had been but also more abstract. The issues, I thought when I wrote 'Universals and Family Resemblances' and 'Principia Metaphysica', were logical and not biological—as Wittgenstein's recurrent talk of 'natural history' seemed to suggest. I have since learned from Peirce and James that it is possible to achieve great clarity and wide scope without abandoning the recognition that to describe thinking and reasoning is to describe thinkers and reasoners—'a *very* familiar class of animals'.[4] I have also learned that to be as systematic and far-ranging as I wish to be is not necessarily to lose Wittgenstein's respect for the particular and his suspicion of the generalities of other philosophers, and one's own. Wittgenstein tried to guard against the ill effects of exaggeration by confining himself to remarks, instances and well directed rhetorical questions. He accordingly risked losing some of the benefits of scope and range that more systematic philosophers have purchased at the risk of simplification and distortion. In self-examination and in self-tuition, as in tutorial teaching of others, we combine both sets of advantages if we allow the dangerous exaggerations to display themselves and then correct them by the proper provision and judicious use of question time.

⁴ Wisdom, *Philosophy and Psycho-Analysis* (Oxford: Blackwell, 1953) p. 112.

Notes on Contributors

Renford Bambrough is a Fellow of St John's College, Cambridge. He was Editor of *Philosophy*, 1972–1994. Among his publications are *Reason, Truth and God* and *Moral Scepticism and Moral Knowledge*.

Sophie Botros is a Lecturer in Medical Ethics at King's College, London University. She has published articles in moral philosophy, and she is presently completing a book which approaches the study of moral philosophy by examining ethical issues in medicine.

Nancy Cartwright is Professor of Philosophy at the London School of Economics. She is the author of *How the Laws of Nature Lie* and *Nature's Capacities and their Measurement*.

Stephen Clark is Professor of Philosophy at Liverpool University, and joint editor of the *Journal of Philosophy*. His books include *The Moral Status of Animals* and *How to Think about the Earth*. His new book, *How to Live Forever*, is a study of what science fiction writers have had to say about immortality.

David Cooper is Professor of Philosophy at The University of Durham. He has been a Visiting Professor in the USA, Canada, South Africa, Germany and Malta. The most recent of his books are *Existentialism: a reconstruction*, *A Companion to Aesthetics*, and *World Philosophies: an historical introduction*.

Roger Fellows is a Senior Lecturer in Philosophy at Bradford University and Head of the Department of Interdisciplinary Human Studies. He is co-author of *What Philosophy Does*, and he has written articles and reviews on the philosophy of mind.

Robert Grant is Reader in English Literature at the University of Glasgow and he has lectured extensively in the USA and Eastern Europe. He is the author of *Oakeshott*, and he has also written many articles on literary, political and philosophical topics.

Kwame Gyekye is Professor of Philosophy at the University of Ghana, Legon, Ghana. He is the author of *An Essay on African Thought: The Akan Conceptual Scheme*, and he was a Woodrow Wilson Fellow in Philosophy, 1993–94.

Willem Hackmann is the Senior Assistant Keeper of the Museum of the History of Science in Oxford and a Fellow of Linacre College. He has written *Seek and Strike, Anti-Submarine and the Royal Navy 1914–54*, and *The History of the Frictional Electrical Machine 1600–1850*. His main interest is in the history and philosophy of instrumentation.

Robin Hendry is a Lecturer in Philosophy at Durham University. After studying Chemistry and Philosophy at King's College London, he studied for his doctorate at the London School of Economics. He is presently writing a book on aspects of the philosophy of quantum mechanics.

Michael Smithurst lectures at the University of Southampton. He has also taught at New College, Oxford and the City University of New York. He has published papers in the philosophy of science, applied philosophy and Hume.

Anthony O'Hear is Professor of Philosophy at the University of Bradford. He is currently Director of the Royal Institute of Philosophy and Editor of *Philosophy*. His philosophical books include *The Element of Fire: Science, Art and the Human World*. He is a regular contributor to *Modern Painters*.

Index of Names

Index of Names

Kemp, M., 29n
Kepler, J., 33, 62 & n
Kipling, R., 160, 164, 167, 169
Kirwan, R., 37
Kleist, von, 22
Krieck, E., 163n
Kuhn, T. S., 32n, 51n, 62n, 195
Kundera, M., 16

Laing, E., 126
Lamb, W., 73 & n, 75, 77, 78, 81, 83 & n
Lasch, C., 15
Laudan, L., 59n, 60n, 66, 71 & n
Leeuwenhoek, A. van, 32 & n, 33, 50n
Leibniz, G., 169, 170n
Leplin, J., 66n
Lewis, C. S., 184
Levi, P., 7
Levy, S., 9n
Libes, A., 37
Leiber, J., 85
Lipton, P., 59 & n, 61n
Long, R., 146
Lorber, J., 117n
Lowry, Lord, 100n, 102n, 118
Lyotard, J-F., 13

MacIntyre, A., 11, 14n
Malcolm, N., 25
Malm, H., 103n
Marcuse, H., 11, 12
Marten, M., 32n
Marum, M. van, 36n
Marx, K., 10, 184
Maxwell, J. C., 23
Mazrui, A., 138, 139n
Mazzolini, R. G., 29n
Mbiti, J. S., 122, 123
McCarthy, J., 85
McMahan, J., 108, 110n, 111–115 & n, 116n
Mellor, H., 19 & n
Michelangelo, 145
Mill, J. S., 177, 181
Milton, J., 192

Minsky, M., 10, 12
Monnier, L. Le, 37
Moore, G. E., 26, 200
Mozart, 150, 151
Mumford, L., 159
Musschenbroeeck, P. van, 37
Mustill, Lord, 100n, 102 & n, 118

Nairne, E., 40n
Needham, J., 48 & n
Neumann, von, 79
Newcomen, T., 21–22
Newton, I., 33 & n, 62
Nietzsche, F. W., 17
Nobili, L., 40n, 42, 43, 50
Nollet, Abb J., 40n
Norman, R., 38n
Nozick, R., 186n

O'Hear, A., 4, 143–158
Oakeshott, M., 186n
Oersted, H. C., 42 & n, 45n
Opoku, K. A., 126
Ormell, C., 19n
Ortous, de Mairan, J. J. d', 37
Orwell, G., 175

Pacey, A., 129n
Parekh, P., 14n
Palmer, F., 154
Park, J., 77n
Parrinder, G., 122
Pascal, B., 179
Pauli, W., 71, 72
Pevsner, N., 143
Picasso, P., 155
Pinch, T., 29n
Pinker, S., 174n, 175n
Plant, G., 40 & n
Plato, 146, 168n, 174, 185, 188, 190, 191, 200
Plum, F., 99n
Pois, R. A., 163n
Popper, K., 7, 19 & n, 25, 27, 174
Priestley, J., 23, 50n
Proust, 153